T0326610

The Science Communication Challenge

The Science Communication Challenge

Truth and Disagreement in Democratic Knowledge Societies

Gitte Meyer

ANTHEM PRESS

Anthem Press
An imprint of Wimbledon Publishing Company
www.anthempress.com

This edition first published in UK and USA 2018
by ANTHEM PRESS
75–76 Blackfriars Road, London SE1 8HA, UK
or PO Box 9779, London SW19 7ZG, UK
and
244 Madison Ave #116, New York, NY 10016, USA

© Gitte Meyer 2018

British Library Cataloguing-in-Publication Data
A catalogue record for this book is available from the British Library.

ISBN-13: 978-1-78308-753-2 (Hbk)
ISBN-10: 1-78308-753-6 (Hbk)

This title is also available as an e-book.

Cover image: Titian, *An Allegory of Prudence.*
The National Gallery, London. Presented by Betty and David Koester, 1966.

Ascribed to Titian (Tiziano Vecellio, around 1488–1576), the image of an old man and a
wolf, a mature man and a lion, and a young man and a dog, looking backwards, directly
at the onlooker and forwards, respectively, has been interpreted in many different ways.
It was given its present title in English – *Allegory of Prudence* – because of an inscription
advising onlookers to take heed of past experiences in order to not jeopardize future
events by present decisions. Thus, there is a connection to the classical notion of practical
reasoning or phronesis, executed within the confines of time – not by outside observers –
and drawing on experience from one case to another. Phronesis, though, had an ethical
dimension, which is apparent neither in the painting nor in the notion of prudence.

CONTENTS

SNAPSHOTS

ACKNOWLEDGEMENTS

The work behind this volume has been partly funded by the Danish foundation TrygFonden. Thanks also to all those organizers of conferences, seminars, workshops and other cooperative projects during the most recent decades who, in a spirit of pluralism, have made room for me to develop my thoughts on the science–society relationships in general and on science communication in particular.

Chapter 1

SCIENCE COMMUNICATION IN DEMOCRATIC KNOWLEDGE SOCIETIES

Science communication idea(l)s are also science idea(l)s. They cannot help but be so. Understandings of science communication and the consequent science communication practices are based on assumptions about science and the roles of science and scientists in society. The currently dominant understandings have a built-in aversion to think about and enquire into their underlying assumptions, but it is urgent, this book argues, that we do actually think about and enquire into such basic ideas and that we open them up for inspection, exchange and possible revisions. It is urgent because the mainstream approaches to science communication may serve to inadvertently erode the societal context that facilitated the development of modern science as an intellectual endeavour and without which it may prove increasingly difficult to maintain science in that sense.

Modern science spent significant moments of its infancy in the coffee house atmosphere of the Enlightenment era, in an intellectual climate of commitment to free speech and free enquiry, marked by a vivid engagement with societal issues. A modern public of reasoning citizens, the backbone of any civil society, was beginning to materialize. With their eagerness to exchange opinions and their omnivorous interest in just about everything, they were preparing the ground for the modern democratic institution of public discussion on public affairs. Early modern scientists contributed to, and the development of modern science was nursed and protected by, this liberal and pluralistic intellectual climate. It is a significant component of the luggage of modern science, which could hardly have reached maturity without it. But it is fragile freight, vulnerable in particular to those other elements of historical luggage that originate in religious strife, civil war and a commitment to monistic truth-seeking.

There appear to be no traces of a pluralistic heritage in the dominant science communication paradigm, pursued as a matter of routine by the majority of participants in exchanges on science-related issues. The paradigm focuses on the dissemination of scientific truth-claims but does not know how to deal with disagreement as anything other than disorder, and is impotent when it comes to, or ought to come to, exchanges among different points of view. Suited for the conventional classroom – or, sometimes, the pulpit or the market stall – it is a didactic paradigm in the sense that it is concerned with the communication *of* scientific findings from knowers to non-knowers, rather than with communication *about* scientific enterprises.[1] The circumvention of the latter activity may, however, prove perilous to societies pervaded by science-related public affairs – res publica – and political issues. Scientific truth-claims may end up devouring the political activity of public exchanges among different points of view.

To make room for both of these distinctly different, but also increasingly interrelated activities – scientific enquiry and political activity – we need awareness of the rather messy and to some extent contradictory origins of modern science. Without such appreciation, both kinds of activity might be endangered to the detriment of future generations.

Founded on the crude assumption that science and politics constitute a straightforward dichotomy or dualism, representing Truth (good) versus Power (bad), the kinds of knowledge societies that are currently growing upon us seem unaware of the above interconnections. There is a corresponding unawareness of how short the distance might be between the assumed dualism of *Truth versus Power* and an idea(l) of *Truth as Power* – which, in turn, might even more easily lead to *Power as Truth*. Neither of these assumptions leaves room for discussions from different points of view. Therefore, there is a need to consider how to maintain, or somehow reintroduce, a liberal and pluralistic intellectual climate into exchanges about science-related public affairs and political issues.

The ways we communicate about science-related affairs are crucial to the further development of current knowledge societies as pluralistic, democratic societies with room for civilized disagreement and political discussion. Therefore, it is time to rethink the ways science may be told and talked about. Considering the significance ascribed to science as a founding element of modern, Western civilizations, this is no mean challenge.[2] Few questions go more deeply to the roots of modern societies than the question of how we communicate about science. Nonetheless, the development during the recent decades of a professional field of science communication, accompanied by the growth of public relations (PR) departments at universities and other research institutions, seems to have taken place on the basis of the tacit agreement that science communication is primarily a specialized, (socio)technical task of knowledge dissemination. Focusing on know-how, fundamental questions pertaining to the roles of science in society and to the identity of scientists have been largely left behind.

The rationale behind this volume is different. Viewing science communication as a general rather than a specialized topic, and as a practical-ethical rather than a technical challenge, it enquires into the apparent background assumptions of mainstream understandings of science communication, asking how they may have come about, where they might be taking us and whether it might be possible to progress in another direction. Focusing on contextual aspects of science communication understandings and drawing mostly on old sources – some of them very old, indeed, and rarely present in writings about science communication – the argumentation stands somewhat apart from the current scholarly science communication discourse with its affinity for social-scientific frameworks and approaches. Using different ingredients, I have prepared a different brew. This should not be perceived as a denigration of other approaches any more than the serving of cocoa constitutes a denigration of coffee. Addressing a wide and widely dispersed audience of everyday practitioners and using the lenses of history and philosophy to explore the background of widely diffused practices, the intention is to supplement the general discourse – forming part of a much larger discussion about science in society – with perspectives that have been widely neglected. You might call it a back-to-basics approach; only, the basics of science communication appear never really to have been attended to.

Truth and Disagreement

Current knowledge societies have come into being through the expansion of scientific methods and frameworks of thought to evermore areas of life and, based on an understanding of science as an all-purpose problem solver, support its further expansion. That development is less pragmatic and down-to-earth than it may appear at first glance. It comes with a relentless extension of the domain of the logic of universal truth and its technical equivalent – correct solutions. Potentially, it seems, science can provide answers to all questions and solutions to all problems. There is nothing, really, to disagree about. Disagreement appears as no more than a symptom of inadequate knowledge – in those who disagree or because science in that particular field is still immature – or as the result of a clash between irreconcilable moral principles. As a consequence, democratic knowledge societies are challenged as political entities in the classical, pluralist sense, characterized by continuous discussion among different points of view and ways of reasoning and using disagreement as a vehicle for discussions, deliberations, negotiations and compromises from one case to another.

The didactic science communication paradigm of science dissemination is an offspring of the view of science as an all-purpose problem solver and facilitates the further development of knowledge societies that rely on scientific – or seemingly scientific – solutions to all sorts of problems. At the same time, the paradigm may contribute to the erosion of such societies as political and democratic entities. This might be seen as a science communication dilemma, presenting us with a stark choice between political pluralism and the advancement of science. The apparent dilemma, however, is founded on the presupposition that science and politics are competing activities, concerned with similar questions in different ways. The dilemma disappears if science and politics are taken to be substantially different activities, suited to dealing with different kinds of questions, to be dealt with and spoken about in different ways. The transmission of scientific knowledge and the discussion of science-related political issues, then, come to be seen as different – although frequently interconnected and sometimes conflicting – kinds of activities.

An assumed dichotomy or dualism of science versus politics lies beneath the understanding of science and politics as competing activities. Based on that assumption, there is no substantial difference between the two kinds of activity. Rather, they represent the opposite sides of the same coin. As such, they are mutually exclusive and it is impossible to have it both ways. Each of us will have to choose to side with either science or politics, hoping for one to swallow the other. As both kinds of activity would be destroyed in the process, that would make any science–politics distinctions superfluous.[3]

The enquiry and the argument to be unfolded in the following pages are born out of a concern that humankind might actually lose these two distinct civilizing achievements – modern science and modern, democratic politics. To maintain them, I argue, it is necessary to view them as substantially different activities, representing different logics that are equally valuable but not directly comparable. According to one logic, the logic of science, the notion of *Truth* is pivotal. According to the other logic, the logic of politics in the classical sense – currently the endangered species – the notion of *Disagreement* is pivotal.

Now, insofar as true – or correct – answers can be found to a question, then, of course, there is no place for substantial disagreement with respect to that question. People may disagree about how to identify those true or correct answers, but no more. If all possible questions belonged to that category, then no other logic, no other framework of thought than the logic of science would be needed.

Conversely, it makes no sense to apply the criterion of truth in connection with a question that may be answered in multiple, reasonable ways, none of them truer than the others. If all possible questions belonged to that category,

then it would appear justified that a classical political logic – prescribing deliberation based on exchanges among different points of view – should generally
prevail.

But why would or should only one logic prevail? Focusing on science-
related political issues, I will make the argument that we can and should have
it both ways, deciding from one case to another what approach – or mix of
approaches – seems most suited and, thus, what variety of science communication we should pursue. Different kinds of issues are suited to different
kinds of approach. Some issues or aspects of issues are of a scientific nature,
meaning that there are unequivocal answers and effective solutions to be
found. Other issues or aspects of issues are of a political nature, meaning that
they relate to human affairs and actions, the consequences of which – not
being guided by universal laws – cannot be foretold. When deciding on action,
therefore, humans have to rely on their judgement, taking a multiplicity of
points of view into consideration from one case to another. Scientific questions
should be dealt with by way of scientific enquiry. Political issues should be
dealt with in the first place through exchange among different points of view.
Decisions on how to proceed from one case to another are themselves matters
for discussion.

The argument is pragmatic and – as distinct from the instrumentalism of
American pragmatism[4] – an offspring of the classical, Aristotelian logic of
politics. It does not go along, in other words, with dominant understandings
of politics as either the opposite or the application of science. Politics is not
defined by its assumed similarities or lack of similarities with science but is
viewed as an activity in its own right.

But what do I mean by 'science'? The current use of English as a lingua
franca has caused confusion in regard to terminology. For instance, science, as
a term, when translated directly into the German *Wissenschaft* and its Nordic
relatives – and then back to English again – seems frequently intended to signify just about anything academic. That, however, is not the meaning of the
term here. Instead, science – and science-based approaches – signifies science
in the strict sense. The exact sciences constitute the model.

The exact sciences deal with exact questions and are characterized by their
search for exact, precise, unambiguous and universally valid explanations
of causal connections. Based on empirical studies and quantification, such
explanations may pave the way for technical solutions to technical problems.
There is a demand that scientific evidence leading to scientific knowledge claims
be reproducible. There are assumptions that the objective and subjective, and
the descriptive and normative can – and should – be radically separated.
Although these and related assumptions have been widely disputed, they
have remained pivotal to scientific methodology. Strict science is committed

to pure description, to idea(l)s of value neutrality and to impersonal, outside observation – as opposed to participation – as a marker of objectivity. Taken together, these criteria form the basis of what is frequently referred to as *the* scientific method. They also precondition a license – claimed by scientists and granted by society at large – to make strong knowledge claims about how things really are (or seem to be at the present stage of scientific development). These criteria form the basis of the authority of science as credible, legitimate, trustworthy, realistic and a source of 'reliable and useful predictions'.[5]

As a term, science connotes a body of knowledge and rational method-ology, an intellectual endeavour, a specific logic of enquiry, a particular aca-demic tradition, a societal institution, a collection of scientific disciplines, a community of scientists – and there may be many more such connotations. Importantly, some even appear to identify with science as a belief system or an ideology. I use the term to signify one or several such aspects, specifying when necessary. I do not use it to make any general statements about scientists as individuals.

The sorts of evidence and knowledge that science brings forth concern universal and technical questions. That kind of knowledge accumulates and is transmissible. Because scientific facts are meant to be impersonal and inde-pendent of context, they can be transferred from one place to another and among persons. Their features can be imitated and they can be taught. They are eminently suited to didactic approaches in the sense of dissemination. And science communication has actually for centuries – long before the present terminology evolved – been widely perceived, irrespective of context, as a didactic enterprise with the purpose of transmitting knowledge from knowers to non-knowers.

Didactics presupposes a knowledge deficit in pupils and students. That is the raison d'etre of teaching. From a democratic point of view, however, grave problems arise when public exchanges regarding the steadily increasing number of science-related public affairs are seen as instances of an overall didactic enterprise aimed at a knowledge-deficient general public. The basic problem is threefold.

First, the didactic paradigm, tailored to suit exact sciences, does not cater for political disagreement. Science-related public affairs are often anything but exact, but the didactic paradigm deals with them as exact questions and takes for granted that true or correct answers or solutions can be, or have already been, found. As a consequence, the existence of disagreement comes to be seen as a symptom of ignorance and its substantial aspects can neither be properly expressed nor addressed.

Second, the roles of the citizen (mature) and the pupil (immature) are confused. With the noblest of intentions, citizens may be subjected to patronizing or matronizing exercises that do not appeal to their capacity for independent reasoning nor, indeed, acknowledge a need for such reasoning to take place outside the institutions of science. At the same time, scientists cast in the role of teachers appear as non-citizens and, frequently, are discursively excluded from the general public by means of a terminology that radically separates scientists, as experts, from other citizens perceived as the laity.

Third, scientists are cheated of the opportunity to be confronted with non-scientific ways of reasoning that might contribute to resolving the issues they are struggling with.

The didactic science communication paradigm, thus, indispensable as it is in some contexts, comes with severe limitations in other contexts.[6] As examples of the latter are becoming increasingly frequent it is also becoming increasingly urgent to recognize those limitations and take them into account when science-related public affairs are on the agenda. Science communication deliberations need to include reflections on when didactic approaches to science communication are, or are not, suitable, and why or why not.

A conspicuous absence of substantial ideas of politics has been a continuous feature of science communication discourses. Apparently, the ancient idea of science as 'Universal Light' with the potential to answer all kinds of secular questions[7] – and with it the attendant negation of politics as anything other than either the irrational opposite or the rational application of science – has survived centuries of scientific development and expansion. It is, it seems, the founding assumption of the didactic paradigm as the one and only approach to science communication. It caters for truth, outreach, inclusion and promotion but not for disagreement and exchange among equals.

Because of the expansion of science, science communication has come to be concerned with such a diversity of topics and issues that one single category of science communication, based on one specific logic, is clearly inadequate. In particular, a communication logic that evolved to suit the exact sciences is inadequate in an era when, more often than not, science-related issues concern inexact questions, loaded with normative aspects and tied to thick concepts, descriptive and normative at the same time.[8] In some such current cases, knowledge claims may be tied to the terminology of 'research' rather than 'science', but 'research' appears to be widely ascribed presumed scientific qualities as a non-interpretative, fact-producing activity and to be perceived as an advanced version of science, without any definite portfolio.[9]

Classical political thought offers a supplement to didactic science communication insofar as it is possible to identify true or correct answers or solutions to some, but not all, questions or problems – if, that is, some questions and problems are of a technical-scientific nature while others are of a practical-political nature. The supplement comes in the shape of what has been characterized as the political core activity: exchange among different points of view among citizens who share a capacity for reason.[10]

The distinction between technical-scientific and practical-political questions is not simple and cannot be easily executed from one case to another. In actual practice, it is a very complicated distinction to make, demanding a lot of effort and balancing – only to find that most current science-related public affairs contain elements of both varieties. The boundaries are unlikely to be ever beyond dispute and might frequently overlap. Still, the distinction makes sense; and, more than that, it might be crucial to the further development of present knowledge societies, not in the direction of technocracies but as vivid and pluralistic democracies. It has the merit that it allows substantially different categories of science communication to coexist; causing and forcing each of us to think about our approaches to science communication case by case and, thereby, hampering the automatic, unreflected application of the didactic paradigm.

Didactic approaches are justified in – and sometimes outside – science education. Somebody knows something that would be useful to others and that may actually be transmitted from A to B, to be used, perhaps, as the foundation for further scientific research or as input to decision making. But a space – and not a tiny one – has to be carved out for a category of science communication that gives pride of place to the art of conversation – *dialectics*[11] – and facilitates the exchange among different points of view on issues that science cannot solve.

There is conflict between the two categories insofar as both cannot be applied to precisely the same topic at the same time. Otherwise, they are not mutually exclusive but might coexist as complementary approaches. Only, the understanding of science as the universal problem solver, and with it the didactic paradigm, have acquired status as parts of the natural order of things and must be denaturalized – provided with a history of their own, that is – to pave the way for revisions that do justice to current science communication challenges.

I fully recognize that, to many readers, my argumentation may appear rather outlandish. I argue that modernity might still have something to learn from antiquity; that the exact sciences might learn something from the liberal arts, and that the English-spoken science tradition might gain from taking into account understandings that appear in other languages. And I claim that this is relevant to understandings of and approaches to science communication. No doubt, there is room for disagreement, but it is possible, I think, to disagree with my line of reasoning and still find the historical and philosophical perspectives it is connected to useful to reflections and deliberations on science communication that go beyond mere technicalities. However, to make the argument useful to readers, so that they may end up mostly agreeing or mostly disagreeing, some conceptual clarifications must be provided.

My interpretations with respect to the terminology of science have been explained, and readers have been made aware of the fact that I do not use the terms 'politics' and 'political' as terms of abuse. Throughout the book, however, a range of other key concepts, notions and figures of thought appear, a good many of which may be subjected to very different and sometimes even conflicting interpretations. My interpretations of and approaches to such concepts as civilization, pluralism and dualisms, knowledge and intellectual activity are not necessarily the most widely used. Some, moreover, might be put off by my affinity with Aristotelian political thought – a reaction that a few introductory explanations might prevent. To avoid misunderstandings, therefore, and to clarify my position of departure, we now turn to some conceptual reflections and clarifications.

Knowledge Societies as Civil Societies

In his 1994 modern classic on the rise of so-called knowledge societies, German sociologist Nico Stehr found that in such societies 'knowledge, rather than more traditional forms of coercive power, becomes the dominant and preferred means of constraint and control of possible social action'. Although he took care to emphasize that 'knowledge as a capacity for action cannot be reduced to scientific knowledge', his discussions of knowledge societies were almost exclusively discussions of *scientific* knowledge societies in which even critics accepted 'the premise of almost non-existent limits to the influence of science and technology on society' while '[m]ost strategic social, political and economic action' could not 'really afford to bypass science'. Stehr also noted the phenomenon of scientific 'self-objectification', which is currently expressed in, for instance, the tacit demand that even science critique be of a scientific vein.[12]

According to Stehr, 'modern scientific discourse does not have a monolithic quality' and therefore 'becomes a resource of political action for individuals and groups who may pursue rather diverse interests'. At the same time, however, he took for granted that a search for 'elimination of disagreements' is a characteristic of science and, thus, knowledge societies.[13]

Knowledge societies, in brief, are pervaded by science, perceived as a universal problem solver. Are they also civil societies? The answer to that question, of course, depends on the definition of civil societies.

Relatively recently, the notion of civil society has been turned into a sociological concept denoting a sector of society (wo)manned by non-governmental, voluntary associations and separated from the modern state. Presently, most, but not all, who employ this understanding of civil society as a separate sector also separate it from the marketplace. At its point of departure, however, the notion signified a kind of society that depended on – was shaped by – civic activity. Translated from the Greek *polis* – which gave rise to such terms as 'political' and 'polite' – the terms 'civility', 'civilization', 'civic' and 'citizen' all originate in the Latin term for a city or city state.[14] They relate to the living together, and to the conditions for so doing, of a diverse citizenry in a city state. Citizens in antiquity were expected to take part in public deliberations and to carry equal shares of public duties. Citizens in a civil society were peers and had to make room for each other and to take other points of view into account when deliberating on public affairs.

That understanding was still present when, in 1767, Scottish philosopher Adam Ferguson (1723–1816) wrote *An Essay on the History of Civil Society*. Civilization, Ferguson found, depends on a concern in members of the public for 'the general good' – but not, he emphasized, in the shape of 'a propensity to mix with the herd'. Human beings, according to Ferguson, 'when in their rude state, have a great uniformity of manners; but when civilized, they are engaged in a variety of pursuits; they tread on a larger field, and separate to a greater distance', and nothing but 'corruption or slavery' could 'suppress the debates that subsist among men of integrity, who bear an equal part in the administration of the state'.[15]

To deprive 'the citizen of occasions to act as the member of a public' counted to Ferguson as almost a cardinal sin. He specified: '[I]f a growing indifference to objects of a public nature, should prevail, and, under any free constitution, put an end to those disputes of party, and silence that noise of dissension, which generally accompany the exercise of freedom, we may venture to prognosticate corruption to the national manners, as well as remissness to the national spirit.' Linking civic activity and liberty, Ferguson argued that if a nation were given to be 'moulded by a sovereign, as the clay is put into the hands of the potter, this project of bestowing liberty on a people who are actually servile, is, perhaps, of all others, the most difficult, and requires most to be executed in silence, and with the deepest reserve'.[16]

To Ferguson, thus, pluralism and participation in political life were hallmarks of civil societies. In his defence of disagreement he seems to have sided with Aristotle in his ancient strife with Plato on the desirable degree of unity in a society. Aristotle found that excessive unity was likely to degrade city states into mere households (i.e., economies) and thus to undermine their political life.[17] While Ferguson's understanding of a civil society corresponds quite well to my understanding of the notion – not as a space in societies but as a sort of society and a precondition for political democracy – its compatibility with the preceding notion of knowledge societies is more doubtful.

Still, modern knowledge societies and modern civil societies have shared roots in the Enlightenment movements of – roughly – the eighteenth century, and to some there appears to be a rather straightforward correspondence between them. Using the degree of technological development to define the degree of civilization, societies pervaded by science and technology become highly civilized by definition. But are they also civil societies? And, in particular, can a search to eliminate disagreement be made compatible with the diversity of opinions and civic activity of civil and democratic societies?

There is no direct fit. Even knowledge societies provided with democratic institutions may evolve into mere technocracies if they choose collectively to put all their faith in science as a solver of every conceivable kind of problem. On the other hand, there is hardly any iron law of nature that prevents democratic societies from relying on and thriving by the advancement of scientific knowledge, accompanied by technological development, while at the same time maintaining a high level of civic activity, including exchanges between conflicting interpretations of, and approaches to, science-related political issues. Science communication practices may hamper or support such twin commitments. Current mainstream approaches to science communication belong emphatically to the former category. Which is one good reason for them to be rethought.

Distinguishing between knowledge societies and industrial societies,[18] Stehr argued that '[u]npredictability, uncertainty and fragility are much more likely to be salient features of knowledge rather than industrial societies'.[19] Without actually pointing to it, he thereby established a connection between knowledge societies and classical political thought. Aristotle's notion of human life as praxis – including politics as its noblest and most demanding form – was founded precisely on the assumption that uncertainty and unpredictability are conditions that humans cannot circumvent and, therefore, must find ways to cope with. This was the point of departure for his pluralist understandings of politics and, thus, of civil societies.

The Aristotelian statement – that life is action or practice, not production – is crucial to this understanding of politics.[20] The notion of praxis captures an idea of the world of human affairs as a specifically human dimension, belonging to an ontology in three dimensions. The human world differs, it is assumed, from the universal dimension inhabited by gods and marked by the complete absence of limitations and restrictions. It also differs from the general animal kingdom inhabited by non-human animals and marked by nothing but limitations and restrictions. The world of human affairs is marked by limitations and restrictions, but human beings are free – in a specifically human way – because of their capacity for thought, speech and, consequently, reason and action.

Everything human is limited in time, in space and because of the plurality of humankind. The world of human affairs is defined by speech and action or practice and is assumed to be devoid of absolutes and certainties; limitations and restrictions relate not only to time, space and biological needs but also to the fact that there is a plurality of humans, and all have different perspectives on human affairs. The latter fact, however, is not merely a restriction. Combined with the human capacity for speech, it also enables humans to deal with human affairs in a specifically human way – exchange of points of view.

Thus, within the framework of praxis, speech is paramount to human life and facilitates the exercise of practical reasoning, *phronesis*, in political life.[21] Practical reasoning is a worldly, temporal and personal kind of reasoning, suited to the practical conditions of limitations, diversity and uncertainty and concerned, from one case to another, with factual and ethical aspects of the possibilities for action.[22] Thus, it is distinctly different from other forms of reason, from *episteme*, which is concerned with universal truth, and from *techne*, or technical reason, which is connected to the production and control of things and includes the possible use of force.[23]

The open-endedness of human languages, the fact that speech is always open to interpretation, marks speech out as the proper medium for grasping human reality insofar as it is taken to be marked by similar features of uncertainty and diversity and, thus, to be characterized by unpredictability.

It seems a sensible course of action – if Stehr was right that unpredictability and uncertainty are elements of the human condition that are becoming more obvious in knowledge societies – to draw on a political philosophy designed to meet those conditions. My suggestion that a political category of science communication as science discussion be introduced represents an attempt to actually do so. Its two interconnected purposes are to maintain and further develop current knowledge societies as civil and democratic societies and to maintain and further develop science as an intellectual endeavour.

The latter purpose, in particular, is not self-evident. The modern idea of scientific knowledge was born with ambiguity vis-á-vis the human activity of thought. To some extent modern science was founded on a suspicion of thought. At the same time, of course, it was unable to avoid practising it. Somehow, that ambiguity must be faced and dealt with. Civil knowledge societies depend on the activity of thought among citizens. To remain civil, they also depend on the maintenance of science as an intellectual endeavour, capable of critical and thorough thought and open to exchange with others. 'If it should turn out to be true', philosopher Hannah Arendt (1906–1975) wrote in 1958, 'that knowledge (in the modern sense of know-how) and thought have parted company for good, then we would indeed become the helpless slaves, not so much of our machines as of our know-how'.[24] Since then, her warning has only become more imminent.

It remains to be seen whether societies that lack the above features of civil – and, indeed, democratic – societies and that are alienated from science as an intellectual endeavour, but intensely committed to science-based technological development, will be able, in the long run, to maintain science as a body of knowledge and rational methodology and to facilitate scientific breakthroughs. Large-scale experiments are currently being carried out around the globe on the seeming premise that the advancement of scientific knowledge is independent of political culture and that science may flourish even if a public exchange of opinions is non-existent. The Enlightenment era was more sophisticated than that. It was equally concerned with the advancement of knowledge and of politics – twin concerns loaded with tensions that are still with us.

With respect to science communication there is manifest conflict between, on the one hand, those notions of a knowledge-deficient public of laypersons that pervade mainstream approaches and, on the other hand, idea(l)s of political equality among citizens. There is also conflict between idea(l)s of, respectively, political pluralism and scientific monism. There cannot at the same time be many valid answers and one true or correct answer to a question. Political pluralism and scientific monism are, however, not necessarily mutually exclusive. Political pluralism cannot allow any one institution a monopoly on reason and is not compatible with the idea(l) of science as universal problem solver, but it has ample room for science-based argumentation. The conflict is too complicated to constitute a dualism or dichotomy.

Truth versus Falsity – and Different Points of View

The dichotomy or dualism – I use the terms interchangeably – is a forceful key figure of modern thought. At least from medieval scholasticism onwards, it has been used as a general formula for thought. In Europe or the West, at least, dichotomies appear everywhere. They have been and are still applied or, rather, taken for granted in academic literature across language borders, affecting understandings of, for instance, the relationships of science and politics as a science-versus-politics relationship and inspiring polarized – and polarizing – attitudes and habits. During recent decades, the preference for dichotomic frameworks of thought has been increasingly subjected to critique. This volume is no exception. Dualistic ways of thinking have informed the logic of science and contributed to shaping dominant approaches to science communication and, therefore, must be confronted if those approaches are to be rethought.

Dichotomies represent a particular variety of distinctions. They express opposite valuations of things, phenomena or qualities that are taken, at the point of departure, to be substantially similar. Thus, this kind of distinction diverts attention from substantial or qualitative differences.

The two sides of an assumed dichotomy are mutually exclusive and interdependent. Truth versus power, objectivity versus subjectivity, observation versus participation and the spiritual versus the material are significant instances of assumed dichotomies that inform the logic of science. Truth is defined by not being false, objectivity by not being subjective, and vice versa. The scheme originates, I suggest, in the notion of universal truth and the corresponding arch-dichotomy of truth versus untruth, falsity or error. They can, in other words, be seen as outcomes of a monistic understanding of knowledge that, in turn, can be seen as a secular relative of religious monotheism. An antagonistic force has been ascribed to monotheistic religions.[25] In my interpretation, their secular relative shares that feature.

Besides diverting attention from substantial aspects of issues and differences of opinion, thus, dichotomic distinctions may encourage tendencies in science communication to antagonize or to perceive interlocutors as antagonists, leading to polarization and demonization. Both features are unhealthy to science communication along the lines of exchanges among different points of view about shared problems.

In the science communication discourse and in approaches to science communication, therefore, dichotomic distinctions should be used with care and after due consideration. Over the centuries, however, the dichotomy has acquired the appearance, not of a particular figure of thought – with a history of its own – but of a universal, or even natural tool for the making of distinctions in general. That might explain the widespread leaning towards applying dichotomic forms of distinction indiscriminately to all kinds of difference. Characteristically, the development of modern science has been connected to opposition to 'the relics'[26] or 'the tyranny'[27] of antiquity and, in particular, to Aristotelian lines of thought.[28] Even the very notion of modernity makes sense only as the antithesis of everything represented by – or ascribed to – 'the ancients' or 'tradition'.[29] And even the capacity for critical judgement – the very ability, that is, to make substantial distinctions – has been ascribed the quality of being negative and in opposition to something as opposed to being positive towards and supportive of it.[30]

Currently, dichotomies seem to be confused by many with distinctions in general. Some, then – wishing to get rid of dichotomic schemes – have set out on a general assault on the very practice of making distinctions at all.[31] Thereby, however, they end up targeting the very capacity for critical thought that might facilitate the careful use of different kinds of distinction from one case to another, in science communication and in general.

In line with the antagonistic scheme, assumed dichotomies may, time and time again, be subjected to a normative inversion,[32] which reverses the attribution of value but keeps the assumption of a fundamental opposition in place. Thus, while preparing the way for a new school, theory or -ism, a fairly uncomplicated return to the original valuation at a later stage – another re-valuation or, if you like, re-volution – is secured. I have used the notion of normative inversions to facilitate my interpretation of populism as inverted elitism (in Chapter 3).

A significant benefit of classical political thought along Aristotelian lines is that it does not generate dualisms. It is a pluralist framework of thought. There is no intention of conquering the world as a whole. Not concerned with questions of a universal or a technical nature, it leaves room for the notions of truth and correct solutions (episteme and techne) outside the domain of human affairs or praxis, but the notions of truth and correct solutions are perceived to be misplaced in relation to practical-political matters.

A good many current schools of social and political thought profess their adherence to pluralism, and there are multiple interpretations of pluralism around. I have found it safest and most useful to return to Aristotle's concepts of praxis and phronesis and their basic assumptions that humans – as death is the only escape route from the world – cannot escape the worldly conditions, cannot avoid being participants in human affairs, but may refine their reasoning on such affairs by making use of the fact that humans differ from each other and represent different points of view. And luckily, so the assumption goes, humans have the capacity to think, to distinguish among different qualities and to discuss their views. Pluralism, in this version, is, at the same time, an idea of aspects of reality and an ideal of the civilized living together, of a plurality of different citizens who are bound together by equal political responsibilities and a shared capacity for thought, speech and reasoning. Speech is considered a source of knowledge, not in spite of, but because it originates in different and sometimes conflicting perspectives and opinions. Disagreement, according to this interpretation, does not make the world go around – that is a completely different phenomenon – but it does keep the political life of civil societies going.

That understanding of pluralism, thus, does not confine pluralism to purely normative or moral questions – actually it does not operate with the idea of the purely normative. Neither does it assume pluralism to be synonymous with tolerance in the sense of a patient acceptance that there are different groups in society. The diversity of humans is not something to be endured, but is a precondition for practical – as distinct from universal and technical – knowledge and for political practice, and it can be seen as an intellectual virtue to recognize it as such.

The distinctions between universal, technical and practical questions or issues are not dichotomic distinctions but substantial distinctions. As such they prepare the ground for substantially different forms of enquiry and exchange. They are not mutually exclusive. They are much too different for that to make sense. That is true, also, of the notions of the social and the political.

Social and Political Animals

The terms 'social' and 'political' are often used interchangeably or, alternatively, political is used exclusively as a term of abuse whereas social appears with neutral or, more frequently, positive connotations such as community, empathy and intimacy. Correspondingly, the classical characterization of humans as political animals has come to be used mostly to connote cynicism and lust for power, whereas characterizations of humans as social animals appear to have a friendlier ring, emphasizing that humans are mutually interdependent and/or indicating that they are fond of each other's company. I use the terms differently, drawing on the Arendtian distinction between the notions of the social and the political and distinguishing between two complementary perspectives on human affairs – a social perspective and a practical perspective.[33] While the former perspective is standard in social science, the present volume has been informed by the latter perspective.

The social perspective represents a view of humans as one of the animal species that lives in groups. In order to study (other) humans from that perspective, one has to adopt the imagined position of an outside observer. This position facilitates that social groups or categories may be identified by the criterion of homogeneity. Patterns of resemblances and differences become visible. Status and power relations and the degree of distance or intimacy within or among groups come into focus. Furthermore, the objects of study appear to the observer as possible targets of socio-technical intervention aimed at affecting the social relationships or mechanisms of or among groups. The social perspective, thus, can be characterized as a relative of the classical notion of techne, extended and applied to human beings and human affairs.

In general terms, the social perspective directs attention to hierarchies and social (in)equalities, to the (un)fair distribution of goods and to the (un)fair representation of different social groups in various settings. The history of the perspective has been marked by ambivalence and conflict between pessimistic (or dystopian) and optimistic (or utopian) social thinkers and has unfolded within a shared framework of assumed dichotomies such as consensus versus conflict and altruism versus egoism. Some thinkers have assumed an original state of warfare and inequality between humans as social animals. Others have assumed an original state of unity, harmony and equality.

When informing the study of communication, the social perspective facilitates a focus on representation of and relations between speakers. Communication easily comes to be seen as a matter of status and power relations. From this perspective, the obvious questions to pursue with respect to communication are: Who does the talking? Is it done in an exclusive or inclusive way? And, is it likely to spur or prevent social conflict?

The classical characterization of human beings as political animals connects to a practical perspective in the Aristotelian sense and connotes a view of humans as beings who are not merely social animals, living in groups like other species of social animals, but differ from other such animal species because of their capacity for action. That capacity, in turn, is assumed to be preconditioned by their capacities for thought, speech and, thus, reason.[34] Because of these features they are able to engage in exchanges from different points of view that enrich and delimit each other and facilitate assessments of the shared conditions for action.

The assumption of a fundamental equality among humans with respect to the capacity for speech and thought is a presupposition or premise of the logic shaped by the practical perspective.

To modern eyes it is curiously indifferent to social relations. With respect to communication there is a focus on the contents of speeches, not on the relations between speakers. From a practical perspective the obvious questions to pursue with respect to communication are: What is being said? How well is it argued? And, does it contribute to a thorough appraisal of issues?

Those questions deserve increased attention in reflections on science communication. It should be kept in mind, though, that because they attribute a knowledge-generating capacity to the very activity of communication, the questions are at odds with the understanding of science as an all-purpose problem solver. They are connected to the view that discussions carried out by political animals constitute a vital, civic activity of enquiry into practical questions that fall outside the religious and scientific domains of universal truth and/or technical problem solving. Science communication undermines that activity if it does not address these political animals, appealing to their capacity for critical thought and reasonable opinion formation, but takes for granted that science and specialized scientists can provide them with everything they need to know.

From a practical point of view, the concept of communication is not accompanied by the kind of ambivalence that is such a prominent feature of the social perspective. Difficulties may arise, however, from the fact that the practical understanding relies on a distinction between the practical and the technical, which is a matter for discussion in its own right. The distinction was never easy to apply, is not included in the logic of science and, on top of that, has gone out of fashion. Currently, the practical and the technical are widely taken to be synonymous, in both common usage and in academic work. Somewhat paradoxically, the broad notion of the practical – and with it its insistence that uncertainty and unpredictability are conditions of human life – has been swallowed by its much more narrowly defined, estranged relative, the notion of the technical. All things, now, appear to be produced or manufactured or, at least, to be producible and thus, by implication, controllable. Mainstream approaches to science communication, with their focus on the transmission of scientific findings, do nothing to further critical reflection on such assumptions.

Science and Science Communication as Intellectual Activities

The tradition of science, accompanied by the didactic science communication paradigm, is loaded with assumptions and tensions that are rarely confronted. Although probably sharing this feature with a good many other cultural traditions, it is a particular disadvantage of the tradition of science that it has a built-in resistance to concern itself with questions that cannot be grasped by the use of methods from the exact sciences. That resistance is a barrier to overcome if science and science communication are to be maintained as intellectual endeavours with a capacity to facilitate critical discussions – and for self-critical revisions.

If confronted directly with some of those assumptions, many natural and social scientists might find that they do not share them. At the same time, however, they may be going along with methods and everyday practices that are tied to them, logically and/or historically. Mainstream science communication routines can be seen as an example of how tacit assumptions about politics, the public and the science–society relationships have acquired a life of their own and, informing actual routines, may end up becoming self-fulfilling.

When applied to science-related public affairs, with their twin connections to politics and science, a continuation along the lines of the currently dominant assumptions about politics might, as a worst-case scenario, result in the end of politics in the classical sense, with dire consequences for civil democracies and for science as an intellectual endeavour. To prevent this from happening, and to facilitate the development of a diversity of approaches to science communication, such basic assumptions must be addressed directly as, precisely, assumptions that may be consciously adopted or modified on the basis of critical enquiries. For several reasons, such enquiries have to transcend the logic of science and scientific methods that characterize the natural and social sciences. First, one cannot properly inspect one's premises on the basis of those very premises. Second, and no less important, the question of how to communicate about science lacks the qualities of scientific questions proper. It is not exact but subject to multiple reasonable interpretations; and, although experiences of relevance to the possible answers may accumulate, they do not accumulate in the scientific sense. There is no reason to expect that an increasing amount of factual building blocks of knowledge will lead us to a true and correct answer. Thus, while notions such as 'the most recent evidence' are, more often than not, misplaced in this context, awareness of the multiplicity of possible perspectives is paramount.

To inform reflections along such lines I use approaches from practical philosophy, drawing as my main inspiration on Hannah Arendt's practical-political framework of thought with its commitment to pluralism.

'The end of the common world,' said Arendt, 'has come when it is seen only under one aspect and is permitted to present itself in only one perspective.'[35] In spite of all current confessions to diversity, the most recent waves of science and science communication enthusiasm appear to be intensifying a general move in that direction. As a countermeasure, expressed also in my choice of literature, I attempt to contribute to the science communication discourse by raising questions about assumptions – only visible from a certain distance – that seem to be informing widespread understandings and practices and to be tied to one particular among many possible perspectives.

The science communication discourse is international. In our contemporary world, this means that it is mostly English-spoken, strongly influenced by US–American understandings and approaches and, due to cross-cultural export–import activity, marred by translational problems among languages that are not directly compatible. My choice of literature pays tribute to those features by emphasizing English-spoken, including a good many American, sources, while at the same time drawing on literature and using examples from other European language areas. My focus is on Europe and for practical reasons most such examples originate in Northern Europe, in German- and Nordic-speaking societies. It is not the aim, however, to portray those particular cultures as models but to emphasize the fact of European diversity and inspire comparisons among language areas. Informed by different strains of the tradition of enlightenment, Europe has a capacity, expressed in different languages and academic and political cultures, for generating different understandings of what science communication should be taken to mean. But in order for that capacity to unfold, the diversity must be acknowledged.

Logics of science communication are cultural outgrowths, connected to idea(l)s of science and based on answers to questions of purpose (why?), substance (what?), position (from where?) and audience (to whom?). On top of those answers, then, comes the question of modes of operation (how?). The current dominant focus on know-how implies that the answers to the four preceding questions may be more or less taken for granted. In contrast, this volume is almost exclusively preoccupied with the questions of why, what, from where and to whom and with the possible connections between certain assumptions and certain answers to those questions.

Overview

The volume is composed of a handful of essays of a kaleidoscopic nature. Together they present the overall argument, drawing on observations, experiences and writings – modified, expanded, combined, integrated, synthesized – from a lifetime of work, both journalistic and academic, connected to science communication.[36] Using Aristotelian political thought as its frame of comparison, the argument focuses on a selection of background aspects and ambiguities that have a bearing on understandings of and consequent approaches to communication about science-related public affairs and political issues.

At the same time, hopefully, each essayistic chapter presents a consistent argument in its own right and may be used separately by readers with special interest in the topic of individual chapters. To facilitate that kind of use, and because the topics of the chapters are heavily intertwined, readers of the book as a whole will come across some repetitive features.

Chapter 2, 'Science as "Universal Light"', discusses aspects of the history of modern science – its early history in particular – that created a tension between understandings of science as a belief system, an anti-ideological ideology, and as an intellectual endeavour with a capacity for critical, including self-critical, thought and exchange. I make the case that early influences from religious fanaticism and civil warfare among confessions infused the founders of modern science with a dread of conflict – and of enthusiasm as a possible precursor of conflicts – but also with a drive to enthusiastically conquer the world in the name of scientific truth, unambiguous, impersonal, untainted by human interpretations and beyond disagreement. Science communication as a didactic-cum-crusading enterprise, carried out by science enthusiasts, with no petty distinctions being made between teaching and preaching, was and has remained crucial to the purpose of conversion.

Waves of science expansion have, I suggest, been accompanied by science communication enthusiasm along such lines. As science has expanded and has come to concern itself with evermore inexact and ambiguous questions, the understanding of science communication as a didactic enterprise has – even when missionary traits are absent – become increasingly inadequate as the one and only approach. However, other traditions of knowledge might have something to offer. The humanities or liberal arts, conventionally occupied with the kinds of issues science is now increasingly concerned with, include communication norms of a more open and questioning nature – in line, actually, with the seemingly forgotten heirloom of science as an intellectual endeavour: the pluralistic debating climate of the Enlightenment.

Chapter 3, 'The Elusive Concept of the Modern Public', looks into assumptions about and understandings of the public in modern democracies. Particular, and particularly critical and detailed, attention is paid to the view that modern societies are divided into the masses and the elites. The features generally attributed to the former – absence of intellectual capacities and leanings prominent among them – probably originate in strongly non-egalitarian contexts and have, I argue, remained remarkably stable for centuries but have been subjected to different normative evaluations by populists and elitists, respectively. An elitism–populism axis has evolved, composed of condescending assumptions about the general public – influential also in social science – and only allowing movement between its poles. Understandings of and approaches to science communication have become tied to that axis and its founding assumptions. They, in turn, may become self-fulfilling when used as the starting point for communication activities and have likely triggered the idea that science communication should aim to fascinate – bewitch, that is – its audiences and promote scientific rationality by appealing to the presumed irrationality of the lay masses.

Chapter 4, 'The Elusive Concept of Modern Politics', makes the case that dominant understandings of politics use the logic of science as their yardstick and, thus, are characterized by a lack of substantial ideas of politics as an activity in its own right. Politics is seen as the irrational opposite or the rational application of science. The chapter discusses the possible cultural background of that basic understanding, explores various versions of it, looks into the curious phenomenon of anti-political devotion to democracy and makes the case that those phenomena and understandings have informed mainstream science communication idea(l)s.

Different journalistic traditions provide a shortcut to understanding how different science communication paradigms, deriving from interrelated political and academic cultures, may come about. Two frameworks of thought on journalism are presented as models en miniature of wider frameworks or sets of ideas and assumptions about politics, democracy and science. The didactic science communication paradigm, in turn, is characterized as an outcome or a close relative of the reporter tradition of journalism. To expand the array of possible approaches to science communication, I argue, it might be helpful to draw on understandings from other traditions of journalism and, thus, other political and academic cultures.

Chapter 5, 'A Political Category of Science Communication', discusses current science communication challenges relating to science in its capacity as a societal institution, frequently occupied with issues that go far beyond the portfolio of the exact sciences. The didactic paradigm was not cut out for dealing with such challenges, I argue, suggesting that the repertoire of approaches to science communication be expanded with a political category of discussions about science-related public affairs and political issues among citizens – some of whom are scientists – who share responsibility for public affairs and a capacity for reason. Science needs reasonable interlocutors from other walks of society and is – as a body of knowledge and rational methodology and as an intellectual endeavour – more likely to be nurtured than harmed by the disagreements, contradictions, critiques and non-scientific perspectives that would inevitably form part of such discussions. At the same time, the cultivation of habits of discussion along these lines is potentially helpful to civil and democratic knowledge societies struggling to cope with the expansion of science in a reasonable way, steering clear of the pitfalls of populism and technocracy.

My overall argument is theoretical and readers might easily be led astray by the use of specific examples. It would, however, seem strange – in a book so concerned with realism and practice – to completely ignore the value of real-life examples. Therefore, Chapters 2 to 5 include 16 textual snapshots of a column-like nature, appearing as separate entities and making points about such examples. They are meant to serve as illustrations of science communication challenges and their connections with understandings of science that make it difficult to talk about as a human activity, complete with limitations, uncertainties and commitments.

Notes

1   I wish to apologize to readers who have a more nuanced understanding of didactics than the one I rely on here. It is neither meant to be derogatory nor to be read as an attempt to interfere in academic and professional exchanges about didactics. I use a didactics–dialectics distinction to emphasize the difference between, on the one hand, the dissemination of knowledge claims, viewed as or pretended to be educational efforts, and, on the other hand, discussions among different points of view. Some may see such discussions as highly educational, which is fine with me, but that is not how I use the term didactics here.

2   Francis Fukuyama's idea of 'the Mechanism' – science driving the development of modern societies – is a typical example of common understandings of the significance of science to modernity. See Fukuyama, *The End of History and the Last Man*.

3   The increasing and almost inevitable use of the terminology of 'research', replacing the terminology of 'science', may be taken to indicate that the distinction between science and politics is actually becoming blurred. For instance, at a 2012 EU conference – 'Science in Dialogue. Towards a European Model for Responsible Research

and Innovation' – 'research', 'innovation', 'problem-solving' and 'policy-making' tended to be used almost interchangeably to signify a kind of production. See the Danish Ministry of Science, Innovation and Higher Education, 'Science in Dialogue'.

4 Rather than taking on a complementary approach to practice, different from scientific approaches and suited to other kinds of questions, American pragmatism seems to be taking science to be the guide to reality and practice in toto. See my discussion of John Dewey, *The Public and Its Problems*, in Chapter 3. Also Bernard Crick, *The American Science of Politics: Its Origins and Conditions*, and Leon Fink, *Progressive Intellectuals and the Dilemmas of Democratic Commitment*, include critical discussions of American pragmatism.

5 See Thomas F. Gieryn, *Cultural Boundaries of Science: Credibility on the Line*, 1: 'Science often stands metonymically for credibility, for legitimate knowledge, for reliable and useful predictions, for a trustable reality.'

6 The question of how the didactic paradigm might affect science when it is actually applied to exact questions is, of course, highly relevant to reflections on the philosophy of science, but lies outside the scope of the present volume.

7 Thomas Sprat, *History of the Royal Society*, 81.

8 Bernard Williams, *Ethics and the Limits of Philosophy*, discusses thick concepts.

9 Corresponding to the partial and gradual replacement of the terminology of science by the terminology of research, the terminology of data has acquired, it has been argued, an aura of truth, objectivity and accuracy that resembles the aura surrounding the terminology of facts. On the latter point, see danah boyd and Kate Crawford, 'Critical Questions for Big Data', and Stefan Strauss, 'If I Only Knew Now What I Know Then'.

10 Bernard Crick, *In Defence of Politics*, represents one of many possible examples.

11 I use the term 'dialectics' in a non-dualist sense to signify exchanges among a multiplicity of points of view. See J. D. G. Evans, *Aristotle's Concept of Dialectic*.

12 Nico Stehr, *Knowledge Societies*, 168, 98, 65, ix.

13 Ibid., 237, 262. Stehr argues that any search to eliminate disagreement is accompanied by uncertainty.

14 Robert K. Barnhart (ed.), *Dictionary of Etymology*.

15 Adam Ferguson, *An Essay on the History of Civil Society*, Part First Section III (Of the Principles of Union among Mankind, 14–17), Part Fourth Section III (Of the Manners of Polished and Commercial Nations, 166–70), Part First Section IX (Of National Felicity, 49–54).

16 Ibid., Part Fifth Section II (Of the Temporary Efforts and Relaxations of the National Spirit, 185–89), Part Sixth Section IV (Of the Corruption Incident to Polished Nations, continued, 225–31), Part Sixth Section V (Of Corruption, as It Tends to Political Slavery, 231–40).

17 Aristotle, *The Politics*, 1261a10, 1263a40.

18 Stehr's distinction between industrial societies and knowledge societies – seen as *post*-industrial – emphasizes the difference between material and virtual production. Thus, the rise of knowledge societies appears as a break with the past. Knowledge societies might, however, also be seen as *hyper*-industrial, because they are marked by the expansion of industrial methods to encompass virtual production. Based on that interpretation, the rise of knowledge societies represents a continuation of the logic of industrial societies.

19 Stehr, *Knowledge Societies*, 236.

20   Aristotle, *The Politics*, 1254aI.

21   For interpretations of practical reason as *phronesis*, see for instance Hannah Arendt, *The Human Condition*; Hans-Georg Gadamer, *Truth and Method*, 312–24; Alasdair MacIntyre, *After Virtue: A Study in Moral Theory*, and Herbert Schnädelbach, *Vernunft*.

22   The English term 'prudence' does not do justice to the concept of *phronesis* because of its lack of an ethical dimension.

23   Hannah Arendt, 'Kultur und Politik'.

24   Arendt, *The Human Condition*, 3.

25   Jan Assmann, *The Price of Monotheism*.

26   Sprat, *History of the Royal Society*, 121.

27   John B. Bury, *The Idea of Progress: An Inquiry into Its Origin and Growth*, 16.

28   Stephen Toulmin, *Cosmopolis: The Hidden Agenda of Modernity*.

29   For a discussion of the contested concept of modernity, see for instance Peter Osborne, 'Modernity Is a Qualitative, Not a Chronological, Category'.

30   Herbert Marcuse, *One-Dimensional Man*.

31   For possible examples of this, see for instance Michel Callon, 'Some Elements of a Sociology of Translation: Domestication of the Scallops and the Fishermen in St Brieuc Bay', and Bruno Latour, 'On Interobjectivity'.

32   Callon, 'Some Elements of a Sociology of Translation', and Latour, 'On Interobjectivity'.

33   Actually, the distinction between the social and the political, employed by Arendt, is rather commonplace and taken for granted among writers in German.

34   Even slaves, because they were human beings, Aristotle mused, shared the capacity for reason. Aristotle, *The Politics*, 1259b26.

35   Arendt, *The Human Condition*, 58.

36   A recent enquiry into the use of well-being and happiness as scientific concepts provided me with a rather extreme case of science exceeding its limits. That enquiry (Gitte Meyer, *Lykkens kontrollanter: Trivselsmålinger og lykkeproduktion* [The Happiness Controllers: The Measurement of Well-Being and the Production of Happiness]), together with a recent contribution to an essay competition organized by the journal *Public Understanding of Science* (Gitte Meyer, 'In Science Communication, Why Does the Idea of a Public Deficit Always Return?'), triggered the synthesization of my work. I have drawn on both when completing the present volume.

Chapter 2

SCIENCE AS 'UNIVERSAL LIGHT'

After more than five centuries, Albrecht Dürer's painting *Christ among the Doctors* is still likely to have an unsettling effect on most intellectuals. Completed in 1506, it is a symbolic representation of a confrontation between *good*, in the shape of Christ, and *evil*, in the shape of learned doctors with demonic features.[1] The painting can be seen as a birth declaration of the modern strategy of demonization. It is also a stark reminder of the ambivalence towards learning and knowledge that forms part of the early, intertwined histories of modern Western science and modern Western thought in general. Both features are still with us – a tendency to demonize opponents and an ambivalence towards learning and knowledge. Both hamper the ability of contemporary societies to sustain habits of civilized exchanges about science- and technology-related issues.

The secularization brought about by the Reformation included a novel leaning towards the demonization of humans. The Devil, who in earlier centuries had been depicted as fantastic and frightening animal hybrids, acquired human forms and faces. Fear and contempt of, for instance, scholastic doctors could now be expressed by demonic representations of them.

The mental climate that accompanied and brought forth the Reformation was marked by a loathing of the Catholic priesthood and scholastic learning. In the 1660s, scholastic learning – taken, it seems, to encompass most of the arts and letters – appears to have been still considered a prime danger and enemy by the founders of the Royal Society, the parent, if ever there was one, of modern science. They challenged the authority of scholastic learning and – carried along by a movement of science enthusiasm – aspired to take its place. That enterprise has been hugely successful.

Today's widespread and forceful institutions of modern science, however, originating in rebellion against former authorities of learning and knowledge, appear to be very much at a loss when it comes to dealing critically with their own current status as knowledge authorities. The identity of modern science is surrounded by ambivalence and tension. There is ambivalence regarding how to deal with critique and critics. Demonization has remained an option. Some science advocates appear to follow in the footsteps of those early representatives of the movement of science enthusiasm who spoke about modern science in meta-religious terms as 'Universal Light'.[2] Others seem more inclined to simply perceive science as an intellectual activity originating in, and still continuously nurtured by, the intense, encyclopaedic interest in the world that characterized the early Enlightenment era.[3] But then again – there is ambivalence also regarding the understanding of the very notion of the intellectual and its relationships with science. It would, indeed, seem strange to deny modern science the quality of an intellectual activity. Anti-intellectual traits can, however, be rather easily identified in its historical baggage. Along related lines, science can be seen as elitist but is also frequently described as a close relative of democracy.

There are loads of disagreement beneath the surface, affecting how it is possible to speak about science outside, and possibly even inside, the scientific institutions. These institutions, in turn, are increasingly powerful societal institutions that tend to take for granted that their public relations – the ways they relate to the public at large – can be classified in a straightforward way as a didactic task of educating the general public. This is not a new phenomenon. The aim of spreading the light of science has been pursued for centuries. But as scientific methodology has expanded and come to be applied to evermore aspects of life, the muting – brought about by didactic approaches – of substantial disagreement, ambiguities and tensions has become increasingly problematic.

How did this state of affairs evolve and where might it take science and society? It seems timely to address the ambivalence directly, to make it talk and to talk about it. That, then, is the aim of this chapter: to trace some of the origins of some of the current tensions within and relating to science and, thereby, to facilitate forward-looking consideration of how to understand and how to speak about science in society.

Modern Science as a Movement

In the early eighteenth century, according to a relatively recent history of the British Enlightenment, science was 'energetically promoted amongst the public. Initially in London's coffee houses, lecturers began to offer demonstrations with globes, orreries and other instruments displaying the marvels of the clockwork universe, while performing chemical, magnetic, electrical and airpump experiments besides'.[4] A *Spectator* magazine of 1711 looked forward to the time 'when Knowledge, instead of being bound up in Books, and kept in Libraries and retirement, is thus obtruded upon the Publick; when it is canvassed in every Assembly, and exposed upon every Table'.[5] There was a movement of science enthusiasm.

Acknowledging the movement aspect of science may take us some way towards understanding the persistent dominance of the didactic frame. Modern science is, of course, much more than a movement. It is a body of knowledge and rational methodology, maintained globally by millions of scientists, aiming to come up with true, universally correct explanations of and solutions to technical questions and problems. And it is an intellectual endeavour, nursing habits of critical and sceptical thought within scientific specialities. Nevertheless, modern science came into being as a movement, dedicated to faith in science as a cause. Aims of conversion – the desire to move others to share the cause – have been present from the 1660s onwards and have also informed the development of science communication paradigms and practices.

The movement aspect of modern science may be difficult, and to some perhaps even painful, to recognize and cope with. This is because the peculiar modern trait of anti-enthusiastic enthusiasm seems particularly strongly expressed in modern science. Born in the wake of the English civil wars, it became marked, in a roundabout way, by the fear of enthusiasm those religious wars had brought about.[6] In a sense, it began its life as an anti-movement movement of anti-enthusiastic enthusiasts, marked by a highly emotional aversion to emotions. That, in turn, has imbued the long-lived tradition of didactic science communication with significant missionary elements. A teaching–preaching ambiguity appears to have been present from the very beginning and is still with us, as is the strong presence of science enthusiasts in the field of science communication.

In order to get a rough idea of how such ambiguous features may have come about, let us take a brief look at four interconnected phenomena that were highly influential in the British Isles when the Royal Society was founded in the 1660s.

Two of those phenomena relate to experiences with religious enthusiasm and concern, respectively, religious truth-seeking and religious civil war. Two others relate to structural, economic and social changes and concern the early connections of modern science to the sphere of production – with its continuous appetite for new technologies – and to the marketplace and the upcoming and ambitious middle classes that challenged older elites. In varying combinations, these early influences – all of which have bearings on understandings of science communication – have continued to make themselves felt.

Influences from religious truth-seeking and strife

The founders of the Royal Society grew up and matured surrounded by religious crusaders who were convinced that they were in possession of the true

Contagiousness and Obsession

In 1919, the Danish bacteriologist Carl Julius Salomonsen (1847–1924) presented a new diagnosis to the world – dysmorphism. Drawing on the Greek *dysmorphos* – deformed – the term was meant to signify a hitherto unrecognized mental disease of a contagious nature. Those afflicted displayed a predilection for distorted, unnatural and ugly shapes. Their ability to recognize and appreciate forms and proportions had been damaged. As a consequence, they created expressionist works of art.

Salomonsen presented his theory to fellow members of a Danish association for the study of the history of medicine. The association, then, published the theory as a treatise. A highly respected scientist, founder and long-time director of the Danish Serum Institute, Salomonsen has been described as the Nordic pioneer of bacteriology and a great source of inspiration to the young scientists to whom he taught epidemiology. A furious debate followed in the wake of his expansion – drawing on his authority as a scientist – of the concept of contagion from the exact sciences to works of art that did not suit his taste. Fifteen years later, however, Salomonsen's peculiar diagnosis could still be looked up in popular Danish encyclopaedias.

Concerned with contagiousness most of his time, he was probably disposed to spot it everywhere. Bacteriology was a young science, testing its limits and with a capacity for fascinating its practitioners as well as its audiences. Moreover, the terminology of contagion was common among social scientists concerned with mass phenomena, and Swedish psychiatrist Bror Gadelius (1862–1932) had described Expressionism as a symptom of pathological disintegration. So, Salomonsen was in line with the zeitgeist when he called Expressionism a psychopathic movement of art, an epidemic agitation psychosis spreading by way of mental contagion, akin to such earlier phenomena as the children's crusades and the self-torturers – brought about, Salomonsen noted, by fanatic agitators.

The textual snapshot about contagiousness and obsession has drawn on Gitte Meyer, *Hjernen og eftertanken* [Brains and reflections], 38–40. Gabriel Tarde (1843–1904) and Scipio Sighele (1868–1913) are two examples of European social scientists who in the late nineteenth and the early twentieth century relied heavily on the terminology of social contagion. See Gabriel Tarde, *The Laws of Imitation*, first published in French in 1890 and Scipio Sighele, *Psychologie des Auflaufs und der Massenverbrechen*, first published in Italian in 1893.

faith and felt obliged to impose that faith on the whole of society. That was the order of the day in the wake of the English civil wars (1640–60) and during Cromwell's Commonwealth regime. Conviction was put in the centre of political conflict.[7] No petty distinctions appear to have been made between religion and politics.[8] Religious truth-seeking was secularized – taken into the course of time[9] – and gave birth to millenarian, chiliastic beliefs about the return of Christ to the planet. As time went by, such beliefs morphed into visions of using science to create a whole new world as well as new humans.

Probably, thus, a monistic understanding of knowledge came naturally to the founders of the Royal Society, many of whom had held high posts in Cromwellian Oxford.[10] They began their shared activity in the so-called Invisible College in 1645 and achieved their royal charter in 1662 as a permanent institution to promote experiments in physics and mathematics.[11]

The new science, like monotheistic religious convictions, became tied to a search for The Truth, preparing the new approach to knowledge for a possible future career as a competing belief system. At the same time, however, religion was – temporarily at least – provided with a domain of its own. Science was linked to material reality as opposed to a spiritual dimension.[12] It was perfectly possible – and, indeed, normal – to be both religious and a follower of the new movement of science enthusiasm. Among radical enlighteners of the following century, however, there was much hostility, at least towards established religion.[13] An assumed dichotomy of science versus religion has been an element of Western discourse ever since. This conflict may be taken as evidence that science and religion have next to nothing in common. Alternatively, it may be taken to indicate that science and religion were competing for the same terrain and, thus, that science enthusiasts – to their possible embarrassment – had a good deal in common with religious enthusiasts.

Writing the first history of the Royal Society, Thomas Sprat (1635–1713) can be regarded as one of the first propagandists of science as a cause. Thus, he is an early representative of the movement of science enthusiasm, marked – like the religious movements of the period – by a striving for purity, conversion and expansion and, thus, by a potential for schism, polarization and fear of possible heretics.

The conflict of 1640–60, it has been argued, 'by polarizing the nation, bequeathed habits of polarized thinking'.[14] Sprat's history is also an early example of science communication of a missionary vein, written by a preacher who perceived himself to be merely a teacher and seems to have been unaware of his own tendencies to polarize.

Sprat's writings were quite fiery. At the same time, they were informed by a fear of conflict. That fear of conflict, a founding element of the logic of modern science, can be seen as another long-term consequence of the fact that the logic – as well as influential strains of modern political thought – was shaped in the aftermath of violent and bloody conflicts, involving a large part of the total population, among fanatics. It has been estimated that more than one in ten of the adult male population bore arms, and that the conflicts resulted in the death of a larger proportion of the population than the Great War of 1914–18.[15] Even the decades after the Glorious Revolution of 1688 were marked by the prosecution of Catholics – so-called Popists – and by continued conflicts among various Protestant sects[16] and followers of differing shades of pantheism, often referred to as radical enlighteners.[17]

It does not take much imagination to understand that this sort of mental climate generated a general fear of conflict and that to the founders of the Royal Society to create a refuge for their scientific activity, protected from the dangerous sphere of conflicting confessions, became a purpose in its own right: 'Their first purpose was no more then [sic] only the satisfaction of breathing a freer air, and of conversing in quiet with another,' Sprat wrote, 'without being ingag'd in the passions, and madness of that dismal Age.'[18]

Against the background of experiences with sectarian violence carried out by fanatics in the name of truth, Sprat was equally convinced that the new science was a source of universal light and truth; the 'true Remedy', redeeming 'the minds of Men, from obscurity, uncertainty, and bondage'.[19] He was also convinced that a belief in scientific truth was different from beliefs in religious truths. Based on pure observation as opposed to enthusiastic participation by potentially fanatical individuals and groups, scientific truth qualified as a representative of light as opposed to darkness and of consensus and unity as opposed to conflict and division.

Accordingly, the aim of searching for the truth was combined with a search for unity and consensus. Human social relations, at least insofar as they included attitudes, opinions, judgements – which might easily diverge from each other and, thus, result in conflict – came to be distrusted. Technical issues, concerning natural mechanisms, were taken to be safe.

When founding the Royal Society, the participants, according to Sprat, did not meet to discuss '*civil business*, and the distresses of their Country' and did not concern themselves with 'politicks, morality and oratory'.[20] While, he argued 'the consideration of *Men*, and *humane affairs* may affect us with a thousand various disquiets', the contemplation of nature 'never separates us into mortal Factions; that gives us room to differ, without animosity; and permits us to raise contrary imaginations upon it, without any danger of a *Civil War*'.[21]

It is certainly not a new and original observation that disappointing experiences with social and political life – few experiences can be more disappointing in that respect than the experience of civil war – may lead people to turn their backs on civic engagement and activity. This point has been made frequently. Referring to the Reformation, for instance, the case has been made that 'the promised simplicities and stabilities of the new evangelical religion offer[ed] the one point of repair from, and contrast to, an otherwise wholly untrustworthy and mutable civic experience'.[22] The advent of modern science offered another such opportunity at a point in time when it was much longed for. It did not, however, prove to be quite as peaceful as was intended. Typically, Sprat – although wishing to preserve peace – could not help finding himself at war. All civil nations should, he wrote, join the armies in a 'philosophical war' against the 'powerful and barbarous Foes' of '*Ignorance*, and *False Opinions*'.[23]

It seems plausible that the development of ideas of science were informed by the very mental climate they were actually intended to counteract. In more than one sense, confessional features, connected to the notion of universal truth, seem to have been mimed. Prominent among such features was the commitment to a monistic search for truth, accompanied by a willingness to carry out crusades in its name. Apparently, tendencies to think in stark terms of pro- versus anti-science attitudes – and to practise science communication accordingly – are present offspring of that commitment. Likewise, the fear of conflict, expressed as an aversion to attitudes, opinions and judgements, has left its traces in the shape of the widespread idea(l) of simply communicating the facts – keeping a safe distance to differing interpretations that might reveal substantial disagreement and generate conflict.

Anti-enthusiastic enthusiasm

The evolvement of the baffling trait of anti-enthusiastic enthusiasm was an overall (and long-lasting) outcome, difficult to deal with and hold in check because it was (and has continued to be) widely unacknowledged. It was recognizable already by 1667 when Sprat enthusiastically reported that young men were now being 'armed against all the inchantments of *Enthusiasm*'.[24]

Enthusiasm had acquired the quality of a term of abuse. In eighteenth-century England, it has been noted: 'Above all things, enthusiasm was regarded with horror, though it is fair to say that enthusiasm was then identified with fanaticism. On an eighteenth-century tombstone was inscribed as the highest of praise, "pious without enthusiasm." '[25] Later, not least during periods of religious revival, enthusiasm acquired renewed respectability, unalloyed by the ambiguity that characterized the decades immediately after the civil wars. Those decades, however, are likely to have left a lasting mark on the logic of science, its methodological approaches, conceptual understandings and communication practices.

To no great avail, the dread of enthusiasm was criticized from an early stage. In a letter concerning enthusiasm, Anthony Ashley Cooper (1671–1713),

Third Earl of Shaftesbury, even ridiculed the idea of anti-enthusiasm. It was, he found, completely impracticable. One might as well attempt to outlaw love. It could not be done. And it should not be attempted. In the absence of enthusiasm all human enterprise would come to a standstill.

Following an interpretation of what Shaftesbury saw as the debating culture of antiquity, he went on to a critique of his contemporaries:

> Not only the Visionarys and Enthusiasts of all kinds were tolerated, your Lordship knows, by the Antients; but on the other side, Philosophy had as free a course, and was permitted as a Ballance against Superstition. And whilst some Sects, such as the Pythagorean and latter Platonick, join'd in with the Superstition and Enthusiasm of the Times; the Epicurean, the Academick, and others, were allow'd to use all the Force of Wit and Raillery against it. And thus matters were happily balanc'd; Reason had fair Play; Learning and Science flourish'd. Wonderful was the Harmony and Temper which arose from all these Contrarietys. Thus Superstition and Enthusiasm were mildly treated; and being let alone, they never rag'd to that degree as to occasion Bloodshed, Wars, Persecutions and Devastations in the World. But a new sort of Policy, which extends itself to another World, and considers the future Lives and Happiness of Men rather than the present, has made us leap the Bounds of natural Humanity; and out of a supernatural Charity, has taught us the way of plaguing one another most devoutly. It has rais'd an Antipathy which no temporal Interest cou'd ever do; and entail'd upon us a mutual Hatred to all Eternity. And now Uniformity in Opinion (a hopeful Project!) is look'd on as the only Expedient against this Evil.[26]

Idea(l)s of complete harmony were, to Shaftesbury, out of touch with reality and would only serve to inspire the kind of virulent conflicts they were meant to prevent. Human life, according to him, was fraught with ambiguities that had to be dealt with and could not be circumvented. He made the point directly and expressed it in his style of writing, marked by an affinity for the combination of contrasts appearing in sentences such as 'the Harmony and Temper which arose from all these Contrarietys' and 'plaguing one another most devoutly'.

It was the overall message of Shaftesbury's letter that whereas excessive enthusiasm was a problem, moderate enthusiasm was a necessity. However, he aired his annoyance with polarizing habits of thought in vain and did not manage to really influence mainstream thought.

The notion of impersonal, objective scientific knowledge, pure of human commitments and interpretations, gained an increasing number of enthusiastic followers, perceived as an idea(l) of knowledge that admitted only the slightest possible room for substantial disagreement and, thus, for violent conflicts. Only two sets of questions were considered legitimate and justified topics for exchange among scientists: What do we see, and what can we measure, as outside observers? And, are the measurements carried out correctly? Today, probably, Shaftesbury would have been amused to see how enthusiasm – now and again rather immoderate – finds ways to express itself in the rhetorical dress, designed to serve as a straitjacket for enthusiasts.

In practice, neither the origins of the feature of anti-enthusiastic enthusiasm nor its possible influences on understandings of science and science communication have been effectively confronted, either in public or within the scientific community. Therefore, it may have maintained much of its original capacity to generate excesses. Those who believe themselves to be completely opposed to enthusiasm are poorly equipped to identify and attempt to moderate that quality in themselves. Enthusiasm, then, may carry them away.

Throughout the history of modernity, now and again, the commitment to science as a cause has actually shown itself in radically utopian fantasies, inspired by such a runaway enthusiasm that is relatively easy to identify by those who are not thus disposed but potentially confusing to those who think of science as the epitome of realism.

A rich source of illustrations of that point can be found in the report from a symposium, organized in London in November 1962 by the then CIBA Foundation. More than 20 highly respected scientists, predominantly biologists from the English-speaking world, gathered to discuss the prospects of modern biology.[27] Their exchanges were fanciful. Participants, for instance, aired science fantasies that included, among many other oddities, visions of producing aseptic humans fitted for life in outer space, the colonization of which by humankind was expected to be imminent. Against that background, a debate took place about the possible future tensions between a new aseptic upper class, well suited for space travel, and old-fashioned humans, stigmatized by badly smelling faeces.[28]

Other participants envisioned the cloning of persons 'of attested ability', allowing them to bring up their own clones. Such cloning schemes, it was argued, would dramatically raise the possibilities of human achievement:

> For exceptional people commonly have unhappy childhoods, as their parents, teachers, and contemporaries try to force them to conform to ordinary standards. Many are permanently deformed by the traumatic

experiences of their childhoods. Probably a great mathematician, poet, or painter could most usefully spend his life from 55 years on in educating his or her own clonal offspring so that they avoided at least some of the frustrations of their original.[29]

Science was presented and discussed as a limitless enterprise: 'You may wish for anything: a cure-all for cancer, a mastery of mutation, an understanding of hormone action, or a cure for any of the diseases you have especially in mind. None of your wishes need remain unfulfilled, once we have penetrated deep enough into the foundations of life. This is the real promise of medicine.'[30]

Other expressions of excessive beliefs in science are more subtle and more difficult to put one's finger on. One example is the gradual naturalization of the idea of science as universal light, as a quiet and seemingly pragmatic, even profane everyday understanding of science as an all-purpose problem solver. The argument that decisions should be 'made in the light of an adequate understanding of the issues',[31] appears to be no more than common sense but may, nevertheless, be based on a deeply rooted belief that science, epitomizing reason and realism, is capable of providing all necessary understanding, has no limits and is somehow beyond assumptions, beliefs and personal judgements.

Belief and scepticism

Such deeply rooted beliefs are not easily combined with scepticism, but the quality of scepticism is widely regarded and claimed as a scientific quality. There is ambivalence. Scepticism is taken to be essential, but only for insiders. The display of doubt concerning scientific knowledge claims in public is a conflictual issue.

Within science, among scientific peers, the reliability of scientific knowledge claims is generally taken to be secured through the activity of organized scepticism.[32] The dominant science communication paradigm, however, has no place for the exercise of scepticism. Rather, prior to the dissemination of scientific knowledge to audiences of lay outsiders it is supposed to be removed, like a scaffold from a completed building. Which understanding of scepticism lies beneath this seemingly contradictory attitude?

In his history of scepticism from Erasmus (1466–1536) to Spinoza (1632–1677), American historian Richard Popkin (1923–2005) recorded aspects of the concept's troubled history, closely connected to religious strife. In the wake of the Reformation and leading up to the dawn of the Enlightenment, scepticism, he argues, was used as a machine of war in religious conflicts – not unlike, it seems, the ways uncertainty is currently being used in conflicts about science-related political issues. For some thinkers, Popkin observed, 'it was a

Holy War to overcome doubt so that man could be secure in his religious and scientific knowledge', but for others the battle was 'not so much a quest for certainty, as a quest for intellectual stability in which doubt and knowledge could both be accepted'.[33]

In religious terminology, scepticism is considered a menace to religion and the religious. The sceptic, as a doubter, threatens belief and is condemned on par with the nihilist or the cynic. With respect to knowledge claims, on the other hand, it has been noted that '[one] of the most important intellectual trends in early modern Europe was the rise of scepticism of various kinds concerning claims to knowledge'.[34] It has, however, never been clear as to what extent and in what way that rise of scepticism was intended to encompass knowledge claims originating in modern science.

From one position, viewing science as an intellectual enterprise, sceptical questions concerning scientific knowledge claims can be seen as challenges that must be met, allowing doubt and knowledge to coexist. According to this view, the expression – whether in scientific journals or in the public domain – of reasonable doubt about specific knowledge claims forms part of the ongoing discussion.

From another position, identifying with science as a cause, the exercise of scepticism in science may be seen along Cartesian lines, as a quest for certainty, a means to gain certainty and, thus, to overcome doubt and insecurity. When scientists have gained (near) certainty it is time to go and tell the people. Outsiders to the scientific disciplines in which knowledge claims originate are not licensed to express critique and scepticism except in the reduced sense of purely normative opposition to – or fear of – proposed uses of the new knowledge. Apparently without any recognition of double standards, the legitimate exercise of doubt concerning knowledge questions is considered the prerogative of the insiders – the representatives of the scientific disciplines.

Along the latter lines, in recent debates on science-related political issues such as climate change or vaccinations or plant biotechnology, the term 'sceptic' has been widely used as a term of abuse to signify non-believers in scientific knowledge claims concerning future events. It is a kind of usage that brings the scientific community dangerously close to an identity as believers even though scientists at large hardly identify with science as a belief system.

'To avoid turning into its opposite' it has been argued 'skepticism must also be skeptical of itself'.[35] Approached as a belief system, however, science is disinclined to take that medicine and to consider its own possible limitations and shortcomings, both in general and from one case to another. The understanding of science as an unlimited enterprise with the capacity to turn on universal light provides scientists and scientific institutions poor protection against the airing of lofty science visions. Moreover, the largely unacknowledged ambivalence towards the exercise of scepticism may trigger the conviction that science, in order to uphold its status as a model of scepticism, should be tough on doubt and on doubters. Such convictions, obviously, hamper open and critical discussions about science-related issues in public.

Influences from economic and social developments

Today's close connections between the world of science, on the one hand, and the sphere of production and the marketplace on the other, go far back in time. British science historian Herbert Butterfield (1900–1979), referred to the end of the seventeenth century when stating:

> The passion to extend the scientific method to every branch of thought was at least equalled by the passion to make science serve the cause of industry and agriculture, and it was accompanied by a sort of technological fervour. Francis Bacon had always laid stress on the immense utilitarian possibilities of science, the advantages beyond all dreams that would come from the control of nature: and it is difficult, even in the early history of the Royal Society, to separate the interest shown in the cause of pure scientific truth from the curiosity in respect of useful inventions on the one part, or the inclination to dabble in fables and freakishness on the other.[36]

Modern science evolved partly to make the world a safer and more prosperous place. However, its development has been marked by continuous struggles regarding the relationship between technical aims of understanding natural mechanisms in order to achieve control – resulting in the application of science and, perhaps, in increased security and prosperity – and epistemic aims of uncovering universal truth(s) out of pure curiosity or veneration for nature.

Although frequently perceived as a dualism, the twin founding concepts of modern science – episteme, related to the search for universal truth, and techne, related to the technical control of objects – were both present from the outset. And, indeed, well-functioning technical solutions, correct answers to technical problems, can be seen as the technical equivalent to the epistemic value of truth. Epistemic value may be ascribed to understandings of natural mechanisms and the development of technical procedures and, actually, even defences of curiosity-driven scientific activity frequently take the shape of references to its possible long-term utilitarian value.

Thus, both concepts – sometimes conceptualized as thought and action (in the reduced sense of technical activity) – may be united in the idea(l) that

Snapshot II

Science as Saviour

In comparisons of the 2008 representation of cancer and cancer treatment in British, Danish and Swedish newspapers – one broadsheet and one tabloid from each country – I looked almost in vain for examples of critical enquiry. Across differences among individual newspapers, newspaper types, and countries, cancer and cancer treatment formed part of a highly symbolic narrative framework that used cancer – 'the Big C' – as a symbol of fate and death, and portrayed science as the road to rescue and salvation. Science was not presented as a human activity, to be questioned and subjected to ongoing scrutiny on a par with other human activities. Rather, it was portrayed as a – or even as *the* – force for good.

Sceptical questions were, as a rule, raised neither about prospective therapies nor about claims about cancer risks, methods of testing or possible preventive measures. Qualifications made by journalists or sources from science were largely absent, but a good many 'miracle cures' were mentioned, some times with eschatological accents: 'Now we are defeating cancer' and 'If the destination is not yet at hand, it is in sight. The endgame has begun.'

While the pharmaceutical industry appeared as no more than background scenery to scientific progress – vested interests were hardly mentioned – politicians rarely appeared, and when they did they were presented as administrators tasked with securing equal access to the latest outcomes of scientific progress for all. Humankind seemed united under the direction of science, in a war against cancer and death, requesting every individual to comply with the lifestyle prescriptions and undergo the cancer tests developed by scientists.

While accommodating celebrations, warnings, behavioural instructions and gruelling reports from the battlefronts, the non-pragmatic framing seems to have hampered critical enquiries into practical questions concerning scientific uncertainties, conflicts of interest and value disagreements.

The textual snapshot about science as saviour is a modified excerpt from Gitte Meyer, 'Fascinating! Popular Science Communication and Literary Science Fiction: The Shared Features of Awe and Fascination and Their Significance to Ideas of Science Fictions as Vehicles for Critical Debate about Scientific Enterprises and Their Ethical Implications'.

science is and should be committed to searching for universal, unequivocal answers and solutions that can be applied regardless of the specific context and substance of issues. Nevertheless, the relationship between epistemic and technological aims was not always easy-going. Thought and action have been taken to constitute a dichotomy. Some, then, have identified science with thought, others with action. This is a long-standing philosophical dispute.

There was and is potential for tension – erupting now and again as direct conflict – between the purposes of, respectively, universal truth-seeking and material progress and between ideals of, respectively, purity and utility. Whereas the former purpose may facilitate understandings of science as a belief system, the latter connects science to the sphere of production and, in market-based societies, to aims of commercial gain. In periods like the present, cultivating material progress and technological innovation as articles of faith, the technical dimension – despised in antiquity but allowed to come into its own as a constitutive element of modern science – not only gains the upper hand but also becomes almost indistinguishable from truth.

Probably from the very beginning of modern science, understandings of science communication have been influenced by both of the founding concepts and the tension between them. Dissemination efforts of a missionary – preaching – nature have been accompanied by efforts to simply transfer know-how and by attempts on behalf of various vested interests to promote specific technologies or technological products.

Evolving simultaneously with the modern marketplace, modern science also evolved alongside those new and upcoming social groups that created and inhabited the marketplace – the ambitious middle classes who challenged the old elites. Possibly, the adoption by early science enthusiasts of an anti-elite identity was an outcome of those developments and can be seen as an act of solidarity with the new groups. The identity was expressed, not least, by efforts to substitute modern science for scholastic learning. It seems plausible that it contributed to the generation of a degree of hostility between the exact sciences and the humanities, seen as the offspring of scholasticism and, thus, tied to the old elites. Because of their interpretative activities, representatives of the humanities could even be seen as speculative and self-serving middlemen, barring – like the Catholic priesthood that the Reformation had revolted against – each individual's direct access to the fountains of truth.

The anti-elite identity, thus, was directed against specific features of the old elites and was coupled with an aspiration to take their place. At the same time, ancient trains of thought, useful to that aspiration, was integrated into the new understanding of learning and knowledge. The idea(l) of fraternities, inherited from the medieval tradition of guilds, was incorporated into the idea(l) of a scientific community,[37] and the notion of the laity, essential to the medieval church and its priesthood, was maintained. Thereby, the foundation was in place for understandings of science communication as relationships between inside members of the fraternity and outside members of the laity, the former transferring pure scientific facts, untouched by human interpretations, to the latter.

The framework matched the needs of a contending elite opposed to the interpretative activities of older elites. It was not shaped to include practices for dealing with the interpretative activities of the contenders as new and victorious elites. Modern science may have been assumed to do away with that kind of activity altogether.

'Things, not words'

An understanding of speech as the main cohesive factor of society can be found in Renaissance writings that take their cues from antiquity, such as the essays of Michel de Montaigne.[38] During the early English Enlightenment, on the other hand, 'things, not words' became a slogan.[39] Anti-rhetorical attitudes, fuelled by animosity to human language and speech, gained momentum.

To some extent, possibly, the slogan may have been provoked by the significance attached to speech in classical thought – connecting appreciation of language with 'the ancients' as opposed to 'the moderns'. Substantially, though, the slogan was tied to an animosity towards interpretation.

The open-endedness and inherent pluralism of human languages are not easily reconciled with scientific aims of reducing complexity and achieving consensus. Neither do those features of speech agree with the normative stance of anti-normativity – the belief that seekers of knowledge can and should avoid making judgements, in particular insofar as they include normative aspects. Human languages are not neutral but interpretative and speakers of human languages are bound to continuously make judgements that often include normative aspects. Thus, there are good reasons why, at an early stage, modern science had to somehow declare itself in opposition to human language and why, in 1664, a language committee of the Royal Society suggested that the English language be disciplined.

Another way of putting it is that influences from all the previously discussed four phenomena came together in animosity towards human languages and speech. The commitment to monistic truth-seeking came with a dislike of open-ended language that left space for different interpretations. The fear of substantial disagreement and conflict was imbued also with a fear of language that might lay bare such potential conflicts. The connections to the sphere of production and the marketplace favoured usage that did not hamper the actual production and dissemination of things. And the alliance with the ambitious middle classes included a shared hostility towards interpretative activities as the main characteristic of the old, scholastic elites. They, in turn, had developed a preference for highly stilted language as a sign of sophistication and as a means to uphold dominance – leanings that were almost begging for opposition.

The language committee, formed by the Royal Society in 1664 to consider how to encourage better use of the English language, found its linguistic ideal in the field of mathematics. The committee recommended that writers should aim to achieve 'a close, naked, natural way of speaking; positive expression; clear senses; a native easinesse, bringing all things as near the Mathematical plainnesse, as they can'.[40]

Sprat was in line with the committee when, in his history of the Royal Society, he contrasted speech with (unequivocal) truth. Devoting a section of his account to 'Their manner of Discourse', he connected speech to social inequality, dominance and power play, conflict and confusion. He criticized 'the luxury and redundance of speech', 'this superfluity of talking', 'the easie vanity of fine speaking', 'this vicious abundance of Phrase, this trick of Metaphors, this volubility of Tongue, which makes so great a noise in the World'.[41]

In a way, Sprat argued, 'eloquence ought to be banish'd out of all civil Societies, as a thing fatal to Peace and good Manners.' Linking 'the Ornaments of speaking' to the passions he asked: 'Who can behold, without indignation, how many mists and uncertainties, these specious Tropes and Figures have brought on our Knowledg?' He then moved on to emphasize the resolution of the Royal Society to 'reject all the amplifications, digressions, and swellings of style: to return back to the primitive purity, and shortness, when men deliver'd so many things, almost in an equal number of words'. As a consequence, according to Sprat, the members of the society preferred 'the language of Artizans, Countrymen, and Merchants, before that, of Wits, or Scholars'.[42]

Human converse could not altogether be abandoned, but Sprat found that speech should be linked closely to things. In effect, it should as far as possible be disconnected from thought, which he contrasted to reality, perceived as no more than the material reality of things. True knowledge, thus, was connected to things as opposed to words.

Such attitudes can be partly traced back to the Reformation with its championship of literal readings of the Biblical scriptures. Contrasted with 'the deceptive veils of medieval allegorical interpretation', literal readings were perceived as common sense and were taken to facilitate the reclaiming of 'interpretive authority from the institution of the Church, whose possessive gatekeepers were academic schoolmen' who had 'selfishly reserved the power of interpretation for themselves, in order to protect their vested interests'.[43]

Against this kind of background, William Tyndale (1494–1536) – a radical Protestant reformer and translator of the Bible into English who was executed by being burnt at the stake – 'preferred explicit discourse to narrative, assertion to dialogue, and the plain, transparent literal sense to the indirections of literary or religious discourse'.[44] Just about one-and-a-half centuries after Tyndale's execution, the recommendations from the language committee of the Royal Society were in line with his demand that all things should be explained 'simply and plainly',[45] and similar preferences were present in Sprat's argumentation. But those preferences were no longer limited to the reading of religious texts. Paving the way for the idea that facts might somehow speak for themselves, Sprat was in favour of literal approaches in general.

The mathematical ideal of language has, of course, had its share of critics. Jonathan Swift (1667–1745) was one of the earliest critics, and a rather harsh one at that. In his narrative of the third travel of Gulliver, Swift ridiculed the aversion to language – and the scientific focus on risk – that Gulliver was confronted with in the country of Balnibari which was governed by scientists who inhabited the flying island of Laputa.

In Balnibari, we are told, several language projects were taking place during Gulliver's visit:

> The first Project was to shorten Discourse by cutting Polysyllables into one, and leaving out Verbs and Participles; because in Reality all things imaginable are but Nouns. The other, was a Scheme for entirely abolishing all Words whatsoever: And this was urged as a great Advantage in Point of Health as well as Brevity. For, it is plain, that every Word we speak is in some Degree a Diminution of our Lungs by Corrosion; and consequently contributes to the shortening of our Lives.[46]

In order to liberate themselves from the need to speak, Swift continues, inhabitants of Balnibari carried large bags, stuffed with things. Meeting each other in the street they would then proceed to show each other things from the bags, thus practising an alternative form of communication that had been liberated from words. But the scheme did not succeed, the narrative goes on: 'And this Invention would certainly have taken Place, to the great Ease as well as Health of the Subject, if the Women in Conjunction with the Vulgar and Illiterate had not threatened to raise a Rebellion, unless they might be allowed the Liberty to speak with their Tongues, after the Manner of their Forefathers: Such constant irreconcileable Enemies to Science are the common People'.[47]

American sociologist Thorstein Veblen (1857–1927) is unlikely to have been amused by Swift's sarcasm. In 1899, more than two centuries after Sprat had formulated his critique of ambiguous language and speech, Veblen, in *The Theory of the Leisure Class*, forcefully repeated basic features of Sprat's argument.

Veblen linked 'the obsolescent habit of speech' – as opposed to 'matter-of-fact-knowledge' and industry – to a 'predatory' leisure class. He faced the same dilemma as Sprat: exchanges by way of human language could not be avoided completely. Advocating 'the use and need of direct and forcible speech', he also opted for a solution along the lines that had been recommended by Sprat and the language committee of the Royal Society.[48]

In this respect at least, Veblen was not alone. Neither were the attitudes to language and speech he represented confined to English-speaking thinkers. In 1895, for instance, French sociologist Gustave le Bon (1841–1931) took for granted that 'illusions and words' were intimately connected.[49] And half a century later, in 1946, British writer George Orwell (1903–1950), argued along somewhat related lines in a critique of stilted language, dead metaphors and standard phrases, which he linked, in particular, to politics. Political language, according to Orwell, was 'designed to make lies sound truthful and murder respectable, and to give an appearance of solidity to pure wind'. Therefore, he called for 'a fresh, vivid, home-made turn of speech' and took a radical separation of language and thought for granted. Thought, he assumed, related to things – with 'pictures or sensations' as possible stand-ins – and was, at the outset, independent from but might be corrupted by words.[50]

Anti-rhetorical attitudes, in short, have for many intertwined reasons been present and have influenced the science tradition from its early days and are likely to have contributed in particular to understandings of science communication as an activity of simply transmitting scientific facts as they are. They are equally likely to have furthered assumptions that scientists were themselves innocent of rhetoric and, thus, that there was no need to critically examine the rhetoric of science, including its forceful rhetoric of numbers.

Anti-intellectualism?

Science is an intellectual endeavour in spite of the fact that anti-intellectual traits are present in the tradition of science and have been noted frequently by historians.[51] The logic and methods of science may have been developed to confine scientific enquiry to impersonal observation, measurement and calculation, but thorough scientists anywhere and at any time transcend those restrictions. They enquire on the basis of certain understandings and assumptions; they make interpretations and execute judgements. Some may be shy about it, and instructions for the writing of scientific articles may have been designed to conceal it, but that does not alter the fact that the activity of

actually doing science goes far beyond the probably extremely rare phenomenon of purely cognitive activity, connoting intellectual activity in the strictly
limited sense of technical rationality and intelligent calculation as it might
be exercised by robots – or aimed at by individual technicians who adopt
such artefacts as their models. Scientific activity is intellectual in a much wider
sense, but pure cognition seems to have been encapsulated as an idea(l) in the
logic of science.

During spells of uninhibited science enthusiasm, the latter feature of the
scientific logic may trigger a sense of alienation in other intellectuals, perceiving scientists as mere technicians and experiencing that their own intellectual approaches are being marginalized. That, in turn, may provoke
estrangement and hostility between, not least, the sciences and the humanities.

Modern intellectuals have, in fact, been described, not least by American
writers, as alienated. A man, literary critic Harold Stearns (1891–1943) argued
in 1921, might be 'a first-rate specialist in a particular field and yet be fundamentally an ignoramus'. The bald fact was, he elaborated, that 'our universities shelter many well-crammed, narrowly disciplined, expert specialists who
by any proper intelligence-rating come perilously near becoming morons'.[52]
Likewise, in 1962, historian Richard Hofstadter (1916–1970) noted with
regret that the American educational system of his time appeared designed to
produce non-intellectual technicians or experts.[53]

Going much further back in time, we encounter once again Jonathan
Swift's Gulliver as one of the very first alienated modern intellectuals, experiencing estrangement when he visits the country of Balnibari and the Island
of Laputa and is confronted with contempt of his own ideas of learning
and knowledge. The rulers, Gulliver reports, had few other interests than
mathematics and displayed no knowledge of human affairs: 'His Majesty
discovered not the least Curiosity to enquire into the Laws, Government,
History, Religion, or Manners of the Countries, where I had been; but
confined his Questions to the state of Mathematicks, and received the
Account I gave him, with great Contempt and Indifference, though often
rouzed by his *Flapper* on each Side.'[54]

Returning to our own time, relatively recent titles, such as the American *The
Last Intellectuals*[55] and *The Closing of the American Mind*[56] and the British *Where
Have All the Intellectuals Gone?*,[57] have all, with different accents, raised concerns
about the estrangement of intellectuals and the marginalization of intellectual
activity in a wide sense. Meanwhile, a few highly respected scientists, such as
French physicist Jean-Marc Lévy-Leblond and Austrian American biochemist
Erwin Chargaff (1905–2002), have struggled to uphold, within the scientific
community and in wider society, a view of science as a close relative of other
intellectual activities.[58]

The debate about science and the intellectual life at large expresses yet another of those tensions that have marred the science–society relationships for centuries. Just like modern democracies are haunted by recurring eruptions of an apparently chronic crisis of democracy, science and the societies it forms part of seem haunted by recurring eruptions of a chronic, but at times dormant antipathy towards intellectual activity that transcends technical rationality. That tendency, as well as the hostility towards science it may generate, can be seen as a late descendant of the ambivalence towards learning and knowledge that Dürer depicted more than five centuries ago.

The occurrence of eruptions may be stimulated or countered by the ways we communicate about science. At the same time, however, the ways we communicate about science are influenced themselves by those historically rooted features that bring about the eruptions in the first place – the conviction of being in possession of no less than universal light, the fear of substantial disagreement and suspicion of human judgement and language. It is a vicious circle, the possible consequences of which are particularly worrying at a time when the use of methodological approaches from the exact sciences has been expanded to enquiries into an increasing number of inexact questions – a process that has been going on for ages.

Waves of Science Enthusiasm

Going back in time once again, in 1949 we find Butterfield arguing that the scientific movement, as a new factor in the seventeenth century, 'immediately began to elbow the other ones away, pushing them from their central position. Indeed, it began immediately to seek control of the rest, as the apostles of the new movement had declared their intention of doing from the very start.'[59]

One wave of expansion gained momentum during the last decades of the nineteenth century. This was noticeable, not least, in the United States in the wake of the American Civil War (1861–65) where it marked the beginning of the Progressive Era[60] and resulted, among other things, in ideas of scientific management and of a science of politics. The Progressives put their faith in science as a universal problem solver and a vehicle for progress, not least with regard to societal and other human affairs. Science, they believed, could and should deliver 'the fulfillment of America's democratic promise'.[61]

The enthusiasm that accompanied or carried the wave along was still in force, it appears, when American historian John B. Bury (1861–1927), writing in the 1920s, described the second half of the nineteenth century as being marked by 'rapidly growing demand (especially in England) for books and lectures, making the results of science accessible and interesting to the lay public'. This 'popular literature', Bury found, was 'subtly flushing the imaginations of men with the consciousness that they were living in an era which, in itself vastly superior to any age of the past, need be burdened by no fear of decline or catastrophe, but trusting in the boundless resources of science might securely defy fate'.[62]

In the United Kingdom, the belief in those boundless resources was expressed vigorously in 31 issues of a popular science magazine – *The Science of Life* – written and published in 1929 and 1930 by science fiction writer and publicist H. G. Wells (1866–1946), his son, zoologist G. P. Wells (1901–1981) and biologist Julian Huxley (1887–1975). The authors praised 'the rigour of the scientific attitude of mind' as opposed to 'loose and tolerant ideas'[63] and argued that ' "Impossible" is a word scientific men should never use. "Highly improbable" is as far as they are ever justified in going'.[64]

There was a millenarian touch to the hopes expressed by the authors of the popular magazine. They prophesized that '[a]t the end of our vista of the progressive mental development of mankind stands the promise of Man, consciously controlling his own destiny and the destinies of all life upon this planet'.[65] The 'progressive development of the scientific mind,' according to these science popularizers, 'may survive all the blundering wars, social disorganization, misconceptions and suppressions that still seem to lie before mankind. Until in due course the heir comes to full strength and takes possession. But he will survive only on one condition, and that is that he must take control not only of his own destinies but of the whole of life.'[66]

Science communication, in that guise, was far from being limited to the dissemination of dry facts from a reliable body of knowledge. Science was promoted as the source of the ultimate liberation of humankind and as an ideology[67] or a belief system rather than as an intellectual endeavour dependent on critical and sceptical exchange. With blissful insensitivity to the fact that the drawing of conclusions from 'is' to 'ought' are condemned as 'the naturalistic fallacy' within the logic of science, teaching and preaching went hand-in-hand – an inclination that the most recent wave of science and science communication enthusiasm has done little to reverse.

The great awakening of the 1960s

The student movements that gradually materialized in the United States in the early 1960s – spreading to Europe during the following decade – marked the beginning of another wave of science enthusiasm and expansion. Fuelled by, among many other things, despair over racial segregation and violence and, in particular, the Vietnam War (1959/1964–75), the student movements were no less eager to revive American democratic ideals and values than had been the Progressives. Together, the student movements constituted,

Snapshot III

Genetics and Eschatology

British and Danish newspapers of the early 1990s abounded with gene therapy enthusiasm. According to British newspapers, for instance, gene therapy constituted 'the treatment of the future', a move 'into the future' and into 'a new type of society', 'a different species'. This 'Fourth revolution of Medicine' would 'wipe out disease'; it was 'a weapon to change the world' and would facilitate 'total control of reproduction' and 'increase intelligence'. In 'the era of gene therapy', 'life could begin at 100'; or at least there would be 'an average life expectancy of 100 years, thanks to gene therapy eliminating dementia, cancer and Aids'; or gene therapy might even 'expand lifespans to 150 years'. Now that scientists had 'unlock[ed] the key to mortality' and were actually reading 'the Book of Life', it was time to ponder such questions as: 'Could genetic engineering create a master race of children with perfect personalities and features?'

The future was used as an eschatological concept rather than in a straightforward chronological sense. The 'way of the future' related to visions of a 'new world', 'a new type of society', 'ultimate answers' and 'breakthroughs' of a 'revolutionary' and 'dramatic' nature. Thus, the future connoted progress towards human perfection that would, from a utopian perspective, liberate humankind from the limitations of the human condition, but might, from a dystopian perspective have apocalyptic consequences. Hopes were generally high in the United Kingdom. Fear competed with hope in Denmark. In both contexts, the development of eugenics appeared to be unavoidable. As gene therapy successes failed to materialize and American health authorities in 1995 warned against over-optimism, some visions even acquired almost hallucinatory traits in both countries. Scientists were now tracing the source of 'eternal youth' and were on the brink of understanding how to block ageing altogether – 'modern miracles' were still about to be produced.

The textual snapshot about genetics and eschatology is a modified excerpt from Gitte Meyer, 'Expectations and Beliefs in Science Communication: Learning from Three European Gene Therapy Discussions of the Early 1990s'.

like the Enlightenment movement of the seventeenth and eighteenth centuries, a movement of many movements. The reception and interpretation of American values and ideals differed from one place to another, and there was a huge amount of internal differences and struggles, but that does not alter the fact that it was essentially an American movement, reviving a deeply rooted belief in the gospel of science and continuing the conviction of the Progressives that the 'inertia of ignorance, superstition, and blind custom could be overcome only by embracing the powers of the scientific method'.[68]

In their early stages, the student movements of the 1960s and 1970s tended to only exhibit such beliefs in a roundabout fashion. There was despair and disillusionment with the ways science was practised, but the movements were not out to delimit science. They were science reformers, criticizing features that hampered its further expansion. The students criticized the scientific establishment because of its ties to big money. They were appalled by the narrowness and cynicism of academic life, including the abuse of 'academic resources to buttress immoral social practice' relating to the arms race and to manipulation in many forms and places.[69]

They even criticized scientific descriptions – in particular within the field of medicine – for failing to grasp the complexity of human beings and human relations. Accordingly, the belief in science took the shape of demands for reforms that would purify and strengthen academic institutions and adjust the scientific method, enabling it to actually grasp the complexity of human beings and their relations. In a curious circular move, making science coil around itself, science was even turned into its own object of seeming outside observation, turning scientific practices into targets for possible science-based interventions.

Attempts were made – and have continued to be made ever since – to extend and modify methods from the exact sciences in order to allow them to somehow include inexact questions and issues. There was no break with the expansive tradition of science. Rather, there was continuity.

With respect to science communication there was even agreement between established and less established scientists. At the November 1962 London symposium on developments in biology and medicine, organised by the CIBA Foundation, Nobel Prize winner Francis Crick (1916–2004) identified the 'great lack of biological knowledge among ordinary people' as an impediment to the progress and application of biological research. The progress Crick had in mind was the introduction of eugenic measures. Biological education was important, he found, because it enabled 'the solutions to be attained with less stress to the social system'. Not all of the 26 other prominent symposium participants were equally keen on eugenics, but Crick's identification of a knowledge deficit in the public gained widespread support. To

'educate people more in biological facts' was described as 'a necessary pre-
liminary to any action'.[70] The 'average man', it was argued, must be taught to
'understand and appreciate the world that scientists have discovered'.[71] That
average man, in turn, was compared unfavourably to the 'better people' who
were taken to be marked by 'creativity, intelligence, and the leaning towards
science'.[72]

Seven years later, the calls for increased science communication efforts
were matched by related calls from a very different and less established
corner of biology. In 1969, two young British scientists, sociologist Hilary
Rose and biologist Steven Rose, published a joint enquiry into the science–
society relationships. Scientific rationality was expanding to evermore areas,
they found, but 'the gulf between the research activities of the scientists and
popular understanding and aspirations' was 'still deep'. Science had become
'esoteric, accessible only to the high priests, and beyond the comprehension
of the laity', and 'the "everyman his own scientist" ideal' was merely a 'rosy'
ideal.[73]

The Roses were concerned that 'an erroneous "image" of scientists or
engineers among the young' seemed to deter young people from studying
science. It was not, they found, 'the procedures of natural science which are at
fault, but its goals'. Against this background, they declared their commitment
to 'goals of creating an open, accessible and man-centred science' and to a
science that was 'effectively planned according to technocratic criteria'.[74]

However much they differed in other respects, representatives of science,
exhibiting a shared enthusiasm on behalf of science, identified a knowledge
deficit in the general public and assumed that with respect to knowledge-
related issues, society is divided into two groups – scientists and the laity.

In short, the student movements of the 1960s and 1970s can be seen as
representative of a scientific awakening – or a reformation, if you like – that
contributed to paving the way for today's knowledge societies. It stimulated the
development of new scientific specialities and disciplines, led to new practices
of doing science on science and revived old tensions between the aims of
expanding further into the world and the desires to remain pure and uncontaminated by that very world.

At the same time, science went social in another sense to what the student
movements had called for and anticipated. The commercialization of even
university-based scientific activities gained momentum and made itself felt,
not least, in the rapidly expanding fields of modern biology and biotechnology,
and information and communication technologies. While the great community many had hoped for did not seem to materialize and the movements
splintered into factions, a good many scientists appeared to be leaving the
ivory tower only to jump into the marketplace. There is a symbolic value in the

fact that the student activist Jerry Rubin, who made his name in the United States in 1968 when he nominated a swine – Pigasus the Immortal – as a candidate for the presidency, finally made his way as a successful businessman.[75]

Links between science and commercial aspirations did not constitute a new phenomenon. There were, as we have seen, ties between science and commerce from the very outset of modern science. In mid-nineteenth-century France, actually, those ties were sufficiently prominent for Jules Verne (1828–1905) – in his long-lost novel, *Paris in the Twentieth Century* – to expose and question them by inventing the company, Enlightenment Promotion Ltd.[76] The ties between science, industry and commerce have, however, been strengthened significantly as part of the most recent expansions of science. These expansions, in turn, have hugely increased the number of scientists, the number and size of universities and other academic institutions and the competition among scientists to gain funding, whether from public or, increasingly, commercial sources.

Burdened with extreme expectations that it would serve as a vehicle for social change, science was also expected to function, in a very direct way, as a motor of economic growth and a source of commercial profits. And scientists were expected to prove their usefulness in these respects by attracting attention as problem solvers and achieve funding for their scientific activities. These developments, in turn, sparked renewed interest in science communication as crusading exercises or as efforts to achieve publicity, but rarely as intellectual activities in their own right.

Another wave of science communication enthusiasm

The current, rather long-lived wave of science communication enthusiasm and the concurrent development of science communication as a professional activity of science dissemination, promotion, outreach and inclusion are outgrowths of the developments that gained momentum during the 1960s and 1970s.

Corresponding in time with the expansion of the field of biotechnology and the growth of digital information and communication technologies, the wave was notable when, in 1985, the Royal Society published a report on the public understanding of science.[77] The publication was followed by a surge of science communication studies that, by and large, served to continue the view of science communication as a fundamentally apolitical genre of popularization with the aim of educating the general public, one way or another.

In practice, the educational aim – the tacit understanding of science communication as a didactic enterprise – has served as an umbrella for a variety of motives, missionary and marketing among them and frequently expanded with democratization motives, based on a view of science as a kind of surrogate politics or on an idea(l) of scientific knowledge as a good that ought to be equally shared by all. All these activities – more often than not coming in hybrid forms – have generally been seen as instances of the transfer of knowledge from knowers to rather unenthusiastic non-knowers. The knower versus non-knower or expert versus layperson dualism has been criticized, but maintaining the understanding of science communication as 'the process through which scientific knowledge spreads' and 'the presentation of science to wider audiences' the critique has not managed to get beneath the basic assumptions of the didactic paradigm.[78]

The purpose of inspiring love of science is so much taken for granted that often it only appears in side remarks: 'It would be great if all readers loved science for science's sake, but they don't.'[79] Typical statements made by scientists in the British media of the 1990s include that the public should be taught that 'science is good for you', that science should be sold as 'fun', and that it should be adopted as a task to show non-scientists that 'science can be interesting and exciting, not just boring and difficult'.[80] Along related lines, it was the overall aim of a 2011 BBC review of its science coverage[81] to increase 'the firepower of BBC Science'.[82]

The incentives have come from the English-speaking world, but have been widely adopted throughout Europe. Science communication has been made the object of increasing political and academic attention as, on the one hand, a straightforward moral obligation to increase the public understanding of and engagement with science and, on the other hand, an area of socio-technical challenges. In some European countries – Denmark and Sweden are two examples – academics employed by universities are now under an obligation by law to disseminate their knowledge. Also, the EU framework programmes for research have been marked by growing budgets for science-in-society issues that tend to deal with science communication and science ethics as separate entities, approaching science communication as mainly a technical challenge.[83] Current European exchanges on science communication are informed by metaphors from the spheres of production and consumption.[84] Construction, consumption, toolboxes and effective communication are examples. Science communication appears as the final unit in a chain of production. Scientific knowledge is seen as a product and a good for possession, distribution and consumption. 'Upstream', scientists produce knowledge to be packaged and transported 'downstream' to non-scientists as potential consumers.

The dominance of the technical perspective is bound to discourage reflection on whether or not, or to what extent, issues should indeed be considered to be technical in the first place. The scientific methods were not cut out to deal critically – nor, indeed, self-critically – with the very expansion of those methods. The view of science communication as a dissemination exercise is equally poorly equipped to deal with questions that go beyond the scientific logic. Taking the role of science and scientists in public life for granted, the scheme is based on the conviction that it is the task of science and scientists in wider society to increase the degree of scientific literacy in the public, to help individuals to get more accurate pictures of the world and to facilitate the implementation of scientific knowledge so that policy decisions may be based on sound science. Insofar as scientific enquiry comes to be seen as a kind of surrogate politics, the aim of including as many citizens as possible in the scientific enterprise may be added, supported by participatory methods of a socio-technical vein. Currently, that understanding of science may be gaining momentum. Tendencies to confuse the political and the socio-technical indicate as much.

In some parts of Europe, it has been argued, the above developments constitute a novel trend that has supplanted practices of critical (not to be confused with hostile) discussions of science and its possible limits and limitations.[85] This can be seen as a loss. Science, in its capacity as an intellectual activity, might actually profit from such critical exchanges and the possibilities they entail for confronting and coping with its inherent tensions.

Not only the tensions but also some remedies for coping with them have roots in the multifarious history of science. Features from religious fanaticism and strife are there. The founders of modern science rebelled against them initially, but could not help imitating them to some extent, thereby paving the way for a view of science as an ideology. But practices of open and free speech and thought, of vivid exchanges in coffee houses and journals and of civic activity in a myriad of associations are there as well. They have been crucial to the development of science as an intellectual endeavour and must be maintained if science itself is to be maintained in that sense.

Some current attempts to understand and define the intellectual have emphasized that intellectuals 'play a socially interpretive role as speakers, writers, or group leaders based on their own advanced learning',[86] or that they are 'not specialists', 'command the vernacular', write for 'the educated reader', have 'profile and presence' and are not 'ignorant of their civilization'.[87] These interpretations, in short, make the intellectual out to be a widely knowledgeable person with the capacity and inclination to participate as such in public life. The contrast to the idea(l) of scientific knowledge as impersonal, specialized and produced by outside observers is remarkable. It is, however, both perfectly feasible and, indeed, widespread to perform scientific work within the latter framework while at the same time appreciating the value of other kinds of intellectual activity and even recognizing a kinship with other intellectuals, such as writers and artists. Problems only arise if science is turned into an object of worship and cultivated as a belief system, unable to recognize – or even tolerate – other kinds of intellectual activity than its own.

Probably, most scientists most of the time see no reason to give thought to such issues. Moreover, they may find it rather far-fetched to even consider the non-scientific paradox of a tradition of human thought and practice – science – which is burdened with a heritage of suspicion towards human thought. They simply wish to get on with their science and to leave science communication – perceived, possibly, as foreign affairs – to others who are more enthusiastic about the presentation of science to non-scientists and who, therefore, might seem suited to serve as liaison officers. In effect, the development of science communication paradigms and practices may have been disproportionately influenced by representatives of the movement of science enthusiasm, which, due to its roots in religious enthusiasm rather than intellectual exchange, is poorly equipped to engage in discussions among different points of view.

Varieties of Knowledge

It seems timely to look for inspiration from traditions of learning and knowledge that evolved to deal with inexact questions. One such logic is linked to the humanities or liberal arts insofar as they are practised by scholars who enquire *as* humans, studying the world from within, as distinct from perceiving humankind as an object of enquiry to be studied from the outside.

Concerned, as the humanities mostly are, with much broader topics, inexact, multifaceted and often marked by clearly normative aspects and undisguised ambiguity, they also use other approaches than the exact sciences. They are not committed to strictly descriptive and explanatory approaches; they are not based on an understanding of knowledge as necessarily impersonal, context-independent, unambiguous, accumulating and solely deriving from empirical enquiry and outside observation. There is no aim to reduce complexity and identify material cause-effect connections that might be useful to technical problem solving. Instead, there are aims of exploring and documenting complexity, making it accessible to reflection and exchange.

Corresponding knowledge claims have by convention been expected to be softer and more open-ended. Being of an interpretative nature, they are not compatible with claims to ultimate authority. They are, on the other hand, compatible with claims to represent authoritative voices that offer valid interpretations to others.

In essence, the distinction between the humanities and the sciences can be traced far back in time.[88] To a large extent, it mirrors the Aristotelian distinction between, on the one hand, the epistemic activity of truth-seeking, and on

Snapshot IV

Stressing Metaphors

Have you had a burnout? How much cultural capital do you possess? Technical and financial metaphors have for a long time contributed to shaping communication about scientific enterprises concerning non-exact topics of an existential and/or political nature. Building links to the sphere of production and the marketplace, the metaphors provide such topics with an air of exactness and make it appear plausible that technical explanations and solutions can be identified and applied.

American medic George M. Beard (1839–1883) was a master of such uses of metaphors. His treatise on *American Nervousness, Its Causes and Consequences*, published in 1881, illustrates how technical and financial metaphors informed not only how he communicated about his work but even how he approached his topic in the first place.

Beard picked most of his metaphors from the front runner technology of his time – electricity. To him, the human brain was a kind of battery. And American nervousness – stress in today's terminology – was an outcome of shortages of electricity. The overall cause was the high degree of civilization in the United States, putting pressure on individuals. Affected individuals had been overcharging their batteries. The cables to their brains could not transport as much electricity as they were trying to use. As a consequence, they went down with a multiplicity of symptoms, bad teeth, headaches, early baldness, depression, indigestion, diabetes and kidney disease among them. They became insolvent, went bankrupt. Understanding the mechanics behind it was the way forward. In this 'new and immense field', Beard prophesized, there was 'room for an army of workers'.

Financial metaphors are still with us. People may not become mentally insolvent today, but many seem to be in lack of social or cultural capital. Electricity metaphors went out of fashion a long time ago. Currently, computer metaphors abound. The difference does not appear to be substantial.

The textual snapshot about stressing metaphors refers to and uses quotations from George M. Beard, *American Nervousness: Its Causes and Consequences: A Supplement to Nervous Exhaustion (Neurasthenia)*, x; and has drawn on Gitte Meyer, *Lykkens kontrollanter: Trivselsmålinger og lykkeproduktion* [The happiness controllers: The measurement of well-being and the production of happiness].

the other, dialectics as enquiry by means of exchange among different points of view. The latter is ascribed a capacity for critique and considered relevant to deliberation.[89]

Importantly, this framework of distinctions does not presuppose that only one form of enquiry might cover all aspects of reality. In conflict with today's widespread assumption that any topic may become a scientific topic if subjected to scientific methods, the distinction between the two activities is related to their different topics – their substance. The different kinds of topic, in turn, are supposed to inform different methodological approaches. Approximating current usage, the distinction corresponds to a division of labour that delegates universal, context-independent questions to science while reserving practical – including ethical and political – issues to the art of conversation: dialectics.

To a large extent, the classical distinction between epistemic activity and dialectics has been carried on in modern distinctions between science and the humanities, and different modern cultures have made different attempts to protect the different varieties of knowing and reasoning – related to different communicative purposes and models – from each other. Separate language areas went their separate ways. During the most recent decades, though, they have come into direct contact in confusing ways that do justice to neither of the protective systems.

Radically separating the exact sciences and the liberal arts, English-speaking cultures draw on a tradition of protecting the exact, modern sciences – including their didactic science communication paradigm – from the messiness and the potential for disagreement of the humanities. Not included in the family of (exact) sciences, the humanities – or, rather, their core subjects – were hardly ever intended to be treated to that communication paradigm. Nevertheless, two developments have drawn it in that direction: the steady expansion of science into the terrain of inexact and practical questions and translational confusion.

In German- and Nordic-speaking areas in the second half of the nineteenth century, the understanding that the humanities were members of the family of *Wissenschaft* (*videnskab, vitenskap, vetenskap* in Danish, Norwegian and Swedish, respectively) gained strength. Aimed at increasing the legitimacy and status of the humanities, this move is likely to have been partly a response to the expansion and increasing influence and status of exact science. The humanities were defined as *Wissenschaften* – *Geisteswissenschaften*[90] – connected to purposes of understanding (*Verstehen*) as distinct from aims of causal explanation (*Erklären*).[91] Because this knowledge system evolved to make room for rather different understandings of learning, knowledge and reasoning, it might presently have something to offer the field of science communication. Translational confusion, however, gets in the way.

The humanities were and are regular Wissenschaften. But they are not sciences. Currently, nevertheless, 'Wissenschaft' (and its relatives in other languages) is widely translated into 'science' and vice versa. And gradually, via series of unreflected translational moves back and forth between language areas, the distinction has been lost between the sciences and the humanities, tied at the outset to their different kinds of topic and connected to different communicative purposes and models. As a possible consequence, both of the protective systems may cease to work, allowing the didactic science communication paradigm to be employed indiscriminately.

As part of that development, other distinctions might be lost as well although they might be useful to reflections on modes of science communication. That includes different understandings of such concepts as interpretation, objectivity and realism.

Interpretation and realism

Interpretations of objectivity and realism are interrelated. One understanding of objectivity seems to be based on the assumption that for something to be real, it has to exist outside the mind. That understanding – first recorded in English in the 1640s[92] – easily develops into the idea that activity of the mind is somehow unreal, may prevent access to reality and result in a lack of realism. The founding fathers of the Royal Society obviously made this understanding of reality and objectivity the foundation for the development of science as a search for universal truth, to be based on direct observation and without contaminating interference from thought, imagery and words.[93]

Another idea of objectivity suggests that private emotions and pre-judgements should not be allowed to direct[94] – or, according to rigid versions of the idea, even influence[95] – assessments, accounts and reports. In itself, this ideal of objectivity – noted in English in the mid-nineteenth century and taken to originate in German-spoken understandings of objectivity[96] – does not exclude thought from reality and, thus, does not take thought and language to be obstacles to understanding reality. Instead, it separates, more or less rigidly, thought and emotion and provides directives for the activities of thought and interpretation. Because it does not outlaw the activity of interpretation, exchanges about inexact and practical questions might benefit from the least rigid versions of this understanding, distinguishing in a non-dualistic way between thought and emotion. This scheme allows scholars and researchers who work with – and, thus, necessarily make interpretations of – inexact topics to make weaker, openly interpretative knowledge claims and, thus, it may also dispose them to be open towards other interpretations.[97]

The long Western tradition of tension and controversy between different understandings of learning, knowledge, reasoning and related concepts can be viewed as so many opportunities for mutual learning and inspiration. It is potentially useful as a source of inspiration for revising science communication idea(l)s. Europe, in particular, is a rich source of diversity that seems, however, very difficult to mine.

An example may help us better understand the difficulties that may disturb attempts to come to grips with and somehow combine understandings that originate in different logics, rooted in different cultures and based on different assumptions. In the 1940s – in the early childhood, that is, of today's communication studies – American sociologist Robert K. Merton (1910–2003) struggled to understand and pinpoint the differences between American 'mass communications research' and what he called 'the European species' of '*Wissenssoziologie*' or 'the sociology of knowledge'. Merton was animated by a wish to combine the best features from both traditions and was a keen observer of their differences. He appears, however, to have been unable or disinclined to actually recognize the basic assumptions of the Europeans he observed. Convinced that for academic work to be serious, it had to be scientific, he took their humanist approaches to be simple mistakes and was amazed by the observations that to the European sociologist of knowledge 'the very term *research technique* has an alien and unfriendly ring' and that the Europeans were prone to declaring that other scholars would have probably ended up with quite different interpretations of the material at hand. The European approaches, he found, were marked by a commitment to 'diversity of interpretation' and 'an aversion to standardizing observational data and the interpretation of the data'. But that did not make sense to American social scientists with their commitment – shared, it seems, by Merton himself – to the achievement of consensus.[98]

In the decades since Merton made his comparisons, the commitment to what he called diversity of interpretation has reached a low. Assumptions and approaches from the logic of science have become generally dominant, also in the field of science communication. Paradoxically, that development has taken place during a period of time when science has come to be urgently in need of interlocutors from other logics and, in particular, from frameworks of thought with traditions of dealing with wide and inexact topics.

Varieties of science communication: Didactics and dialectics

Representatives of the sciences and the humanities, respectively, have, by convention, been entitled to make different kinds of knowledge claims.

Traditionally, science, confined by its point of departure to exact issues – suited to enquiry by outside observers – has been entitled to make strong factual statements regarding knowledge, at the current state of scientific development, of strictly defined topics. There has been an aim of achieving consensus and closure. And science communication has been perceived as a didactic task, with widely unrecognized missionary or commercial aspects, of transporting packages of such knowledge from one group of persons to another.

The humanities, in turn, have been preoccupied, as a rule, with much wider, less clearly defined questions, frequently relating to thick concepts, descriptive and normative at the same time. Such questions cannot be answered unambiguously. They require interpretative activity and are incompatible with a norm of pure description. Therefore, the enquiries cannot be carried out by way of outside observation, and statements about the outcomes have conventionally been required to make room for other interpretations and positions and to take the shape of contributions to exchanges. Communication about humanist scholarship and research, in short, has been perceived, to a large extent, as a dialectical enterprise although with didactic aspects.

It seems very neat this division of communicative practices between, respectively, the exact sciences and the liberal arts. If actually applied, it might take us a long way towards distinguishing between didactic and dialectical science communication – based as it is on a distinction according to which the topic of enquiry determines the choice of methods and communicative approaches from one case to another. In practice and for a variety of reasons, however, currently that kind of distinction appears to be rarely made. Moreover, methods from the exact sciences have hardly ever been confined to exact questions but, spurred by waves of intense science enthusiasm, have been expanded since the mid-seventeenth century.

Public representations of and exchanges about science have been a feature of public life in Europe for centuries. Some of those representations and exchanges have been highly passionate and have been based on a commitment to science as a belief system – a commitment that is part of the luggage of current knowledge societies. Until relatively recently, however, they all had to take place within a non-scientific, societal space. The expansion of scientific approaches to cover most activities and professions was a thing of the future. Moreover, until the early twentieth century, scientific specialization was still sufficiently limited for physician Ernest Rutherford (1871–1939) to remark that 'no physics could be good, unless it could be explained to a barmaid'.[99]

In practice, scientists formed part of wider society; science was practised, taught, preached and advertised in a wider, non-scientific context and was mostly only indirectly concerned with public affairs and political issues. And science was not immune, for good or bad, to influences from wider society. Some of those influences concerned assumptions about the general public and have significantly informed science communication paradigms and practices. They are the topic of Chapter 3, 'The Elusive Concept of the Modern Public'.

Notes

1 Daniel Arasse, *Bildnisse des Teufels*.
2 Thomas Sprat, *History of the Royal Society*, 81.
3 Ehrhard Bahr, ed., *Was ist Aufklärung? Thesen und Definitionen*; Margaret C. Jacob, *The Enlightenment: A Brief History with Documents*; Margaret C. Jacob, *The Radical Enlightenment: Pantheists, Freemasons and Republicans*; and Roy Porter, *Enlightenment: Britain and the Creation of the Modern World* throw light, from different perspectives, on the Enlightenment era.
4 Porter, *Enlightenment*, 142.
5 Quoted in Porter, *Enlightenment*, 194.
6 Irene Coltman, *Private Men and Public Causes: Philosophy and Politics in the English Civil War*.
7 Ibid., 12.
8 Luise Schorn-Schütte, *Konfessionskriege und europäische Expansion: Europa 1500–1648*, and Blair Worden, *The English Civil Wars, 1640–1660*, both point to the intertwinement of religion and politics in the seventeenth century.
9 As a term, secularization originates in the Latin *saeculum*: age, span of time, generation. Robert K. Barnhart, ed., *Dictionary of Etymology*.
10 Worden, *The English Civil Wars, 1640–1660*, 162.
11 Maurice Ashley, *England in the Seventeenth Century*, 156.
12 Hobbes found that 'all that is real is material and what is not material is not real'. Ashley, *England in the Seventeenth Century*, 115–16.
13 Both Jacob, *The Radical Enlightenment*, and John Redwood, *Reason, Ridicule and Religion: The Age of Enlightenment in England 1660–1750*, explore the attitudes among radical enlighteners towards religion. French Julien Offray de la Mettrie (1709–1751), with his 1747 tract, *Man a Machine*, exemplifies early, anti-religious science enthusiasm.
14 Worden, *The English Civil Wars, 1640–1660*, 153.
15 Ibid., 73.
16 As a term, 'Protestant' is used in many different ways. I use it to denote the huge amount of Christian convictions, Calvinist and Lutheran among them, that evolved in the wake of the Reformation, protesting against Catholicism.
17 Jacob, *The Radical Enlightenment*.
18 Sprat, *History of the Royal Society*, 53.
19 Ibid., 58.
20 Ibid., 82. Sprat's italics unless otherwise stated.
21 Ibid., 55–56.
22 James Simpson, *Burning to Read: English Fundamentalism and Its Reformation Opponents*, 162.
23 Sprat, *History of the Royal Society*, 57.
24 Ibid., 53.
25 Dorothy George, *England in Transition: Life and Work in the Eighteenth Century*, 65.

26 Anthony Ashley Cooper, Third Earl of Shaftesbury, 'A Letter concerning Enthusiasm', 13.

27 Gordon Wolstenholme, ed., *Man and His Future*.

28 J. B. S. Haldane, 'Biological Possibilities for the Human Species in the Next Ten Thousand Years', 343.

29 Ibid., 352–53.

30 Albert Szent-Györgyi, 'The Promise of Medical Science', 195. For further examples of excessive science fascination, see for instance Mary Midgley, *Science as Salvation: A Popular Myth and Its Meaning*.

31 Walter Bodmer, 'Public Understanding of Science: The BA, the Royal Society and COPUS'.

32 Robert K. Merton, *Social Theory and Social Structure*.

33 Richard H. Popkin, *The History of Scepticism from Erasmus to Spinoza*, 110. For an interesting discussion of scepticism and modernity, see Odo Marquard, *Skepsis in der Moderne: Philosophische Studien*.

34 Peter Burke, *A Social History of Knowledge: From Gutenberg to Diderot*, 197.

35 Russell Jacoby, *The Last Intellectuals: American Culture in the Age of Academe*, 73.

36 Herbert Butterfield, *The Origins of Modern Science: 1300–1800*, 185.

37 The perception of the men and women of science as a tightly knit community or brotherhood, bound together and strengthened by shared convictions, was described by British writer and public servant C. P. Snow (1905–1980) in his 1959 classic *The Two Cultures and the Scientific Revolution*, 10: '[T]heir attitudes are closer to other scientists than to non-scientists who in religion or politics or class have the same label as themselves.'

38 Michel de Montaigne, *Essais*, vol. II, 8.

39 Burke, *A Social History of Knowledge*; Porter, *Enlightenment*.

40 Quoted by Claire Tomalin, *Samuel Pepys: The Unequalled Self*, 258. The language committee is also mentioned by Cooper, *England in the Seventeenth Century*, 157.

41 Sprat, History of the Royal Society, part 2, 112.

42 Sprat, *History of the Royal Society*, part 2, 113–15.

43 Simpson, *Burning to Read*, 107.

44 Ibid., 280.

45 Ibid., 109.

46 Jonathan Swift, 'The Text of Gulliver's Travels', 158.

47 Ibid.

48 Thorstein Veblen, *The Theory of the Leisure Class*, 260, 264, 265.

49 Gustave le Bon: *The Crowd: A Study of the Popular Mind*, book II, chapter I, section 4.

50 George Orwell, *Politics and the English Language*.

51 See, for instance, Porter, *Enlightenment*, 23; Theodore M. Porter, *Trust in Numbers: The Pursuit of Objectivity in Science and Public Life*, 195; and Gordon S. Wood, *The Radicalism of the American Revolution*, 240, 369.

52 Harold Stearns, *America and the Young Intellectual*, 21 (Stearns's italics), 107, 108.

53 Richard Hofstadter, *Anti-intellectualism in American Life*, 416–26; 51, 428. Also, in the 1940s, American sociologist Robert K. Merton hypothesized that 'bureaucracies provoke a gradual transformation of the alienated intellectual into the a-political technician, whose role is to serve whatever strata happen to be in power'. Merton, *Social Theory and Social Structure*, 268. Merton seems to have not given thought to the possible

kinship between the logics of science and bureaucracy, respectively, but rather to have perceived bureaucracy as a separate force that might damage individual scientists.

54 Swift, 'The Text of Gulliver's Travels', 139. Swift's italics. As to the notion of 'flappers', Swift, 138, describes Gulliver's first impression of the inhabitants, like this: 'Their Heads were all reclined to the Right, or the Left; one of their Eyes turned inward, and the other directly up to the Zenith'. It seems a reasonable guess that Descartes and Newton, respectively, had taken possession of their mental faculties and diverted their attention from what was actually happening in the world around them. To catch their attention, they were provided with a servant who, by way of continuous disturbance – flapping – was obliged to secure a minimum of presence.

55 Jacoby, *The Last Intellectuals*.

56 Allan Bloom, *The Closing of the American Mind*.

57 Frank Furedi, *Where Have All the Intellectuals Gone? Confronting 21st Century Philistinism*.

58 See for instance Jean-Marc Lévy-Leblond, 'About Misunderstandings about Misunderstandings', and Erwin Chargaff, *How Scientific Papers Are Written*. I owe the reference to Lévy-Leblond to an anonymous reviewer.

59 Butterfield, *The Origins of Modern Science: 1300–1800*, 206. For an account of the early development of scientific thought in the European continent, with a particular emphasis on France, see Armand Mattelart, *The Invention of Communication*, 3–25.

60 The Progressive Era stretched from around 1900 or a little earlier and a couple of decades into the twentieth century. There is no really exact dating. For a discussion of this, see for instance, Bernard Crick, *The American Science of Politics: Its Origins and Conditions*, 27, 49.

61 Leon Fink, *Progressive Intellectuals and the Dilemmas of Democratic Commitment*, 8.

62 John B. Bury, *The Idea of Progress: An Inquiry into Its Origin and Growth*, 345–46.

63 H. G. Wells, G. P. Wells and Julian Huxley, *The Science of Life: A Summary of Contemporary Knowledge about Life and Its Possibilities*, vol. 26, 829.

64 Ibid., vol. 30, 940.

65 Ibid., vol. 31, 972.

66 Ibid., 973.

67 I do not use the term 'ideology' in a Marxist sense, tying ideas to class interests, but to denote systems of assumptions and ideas that tend to function as secular religions and to command strong convictions.

68 Fink, *Progressive Intellectuals and the Dilemmas of Democratic Commitment*, 13.

69 *Port Huron Statement*.

70 Wolstenholme, *Man and His Future*, 274, 284, 367.

71 Hermann J. Muller, 'Genetic Progress by Voluntarily Conducted Germinal Choice', 255.

72 Wolstenholme, *Man and His Future*, 290.

73 Hillary Rose and Stephen Rose, *Science and Society*, 253–54. Recently, the 'everyman his own scientist' ideal has been reinvented as an ideal of 'scientific citizenship' – see for instance Ulrike Felt, ed., *O.P.U.S. Optimising Public Understanding of Science and Technology: Final Report*.

74 Rose and Rose, *Science and Society*, 260, 262, 268.

75 Ingrid Gilcher-Holtey, *Die 68er Bewegung: Deutschland, Westeuropa, USA*, 96.

76 Jules Verne, *The Lost Novel: Paris in the Twentieth Century*.

77 The Royal Society, *The Public Understanding of Science*.

78   Stephen Hilgartner, 'The Dominant View of Popularization', represents a rather early critique of the experts versus laypersons dualism. It is also a rather early example of how that kind of critique has tended to remain tied to a didactic understanding of science communication as science dissemination.

79   Rebecca Skloot, 'Under the Skin: A History of the Vaccine Debate Goes Deep but Misses the Drama'.

80   Quoted in Gitte Meyer, 'Expectations and Beliefs in Science Communication: Learning from Three European Gene Therapy Discussions of the Early 1990s'.

81   BBC Trust, 'BBC Trust Review of Impartiality and Accuracy of the BBC's Coverage of Science'.

82   BBC Trust, 'BBC Trust Review of Impartiality and Accuracy of the BBC's Coverage of Science: Follow Up', 6.

83   European Commission, *Integrating Science in Society Issues in Scientific Research: Main Findings of the Study on the Integration of Science and Society Issues in the Sixth Framework Programme*; European Commission, *Mid-Term Assessment: Science and Society Activities 2002–2006*. For a further discussion of this, see Gitte Meyer and Peter Sandøe, 'Going Public: Good Scientific Conduct'.

84   Sharon M. Friedman, Sharon Dunwoody and Carol L. Rogers, eds, *Scientists and Journalists: Reporting Science as News* is an early example of the dominance of the production language in the academic science communication discourse.

85   Konrad Paul Liessmann, *Lob der Grenze: Kritik der politischen Unterscheidungskraft*.

86   Fink, *Progressive Intellectuals and the Dilemmas of Democratic Commitment*, 4.

87   Jacoby, *The Last Intellectuals*, 12, x, xv, x, xvii, 73.

88   See also John D. O'Banion, *Reorienting Rhetoric: The Dialectic of List and Story*.

89   Aristotle, *Retorik*; J. D. G. Evans, *Aristotle's Concept of Dialectic*.

90   Although sharing roots with the English term 'ghost' – see *Duden, Das Herkunftswörterbuch* – the German terms *Geist* and *geistig* cannot be translated directly into English. Though sharing some connotations with ghost, they are probably most frequently used to broadly signify intellectual activity of a non-calculating and non-religious nature. They relate to an understanding of reality that is not merely material and do not correspond to an assumed dualism of the material versus the spiritual, but belong in another logic. *Geisteswissenschaften*, thus, can definitely not be translated into 'ghost sciences'.

91   Herbert Schnädelbach, *Vernunft*.

92   Barnhart, *Dictionary of Etymology*.

93   Porter, *Enlightenment*; Redwood, *Reason, Ridicule and Religion*.

94   *Duden, Das Bedeutungswörterbuch*.

95   A. S. Hornby, ed., *Oxford Advanced Learner's Dictionary of Current English*. Eighth edition.

96   Barnhart, *Dictionary of Etymology*.

97   Weaker claims are made on behalf of Wissenschaft than on behalf of science. The definition of unscientific as 'not scientific, not done in a careful logical way' – see Hornby, *Oxford Advanced Learner's Dictionary of Current English* (fifth edition) – is not matched by the definition of *unwissenschaftlich*, which is a more neutral term and does not imply lack of care or logic. See *Neues Deutsches Wörterbuch*.

98   Merton, *Social Theory and Social Structure*, 493–509. His italics.

99   Eric Hobsbawm, *Age of Extremes: The Short Twentieth Century 1914–1991*, 538.

Chapter 3

THE ELUSIVE CONCEPT OF THE MODERN PUBLIC

Current discussions about science communication and the roles of science in society tend to frame the relationship between (scientific) expert knowledge and (political) democracy as a social issue, or even as a social conflict between scientific experts and so-called ordinary citizens as social groups. As a probably widely unrecognized and unintended consequence of that framing, scientists appear – in their capacity as scientists – to be excluded from the general public, from the citizenry and the civic responsibility that citizenship implies.

In the city states of antiquity, the classical *polis*, slaves and women were excluded from citizenship, but modern democracies, in principle, grant citizenship to all adults. They are all members of the public. Thus, the tendency in the science–society discourse to exclude scientists from the public or to regard them as extraordinary citizens – whatever that might imply – appears as a thought-provoking anomaly: Which assumptions about the general public or citizenry inform the framing that places scientists outside – or even in opposition to – the general public? And, what is understood by citizenship? How, in turn, might such assumptions and understandings have evolved and how do they affect science communication routines and models of thought?

The discursive exclusion of scientists from the general public seems consistent with a view of the general public as a social rather than as a political entity. The concept of the citizenry connotes the public as a political entity, composed of co-responsible citizens. The concept of the masses connotes the public as a social entity or group composed of so-called common men. It also presupposes the existence of elites. It makes sense to exclude scientists from the public insofar as science is considered an elite activity and society is perceived in terms of a division between the masses of common (wo)men and – or versus – the elites. Scientists, then, appear as somehow uncommon – or extraordinary – men and women. Are they really? What capacities are ascribed to the supposed masses that make them common? Should scientists be considered to constitute a modern variety of aristocracy? If so, the relationship between scientists and other citizens is turned into a relationship of rank and status. But that does not tally nicely with the historical identity of modern science.

Nothing much tallies when it comes to those understandings of the modern public that appear to inform widespread science communication practices and the general science–society discourse. The topic is fraught with tensions and contradictory assumptions, some of which may prove self-fulfilling. What goes around comes around. Assumptions about the public – the main topic of this chapter – tend to come with self-fulfilling qualities. Provided with a history of their own, however, such assumptions may be made visible *as* assumptions rather than as parts of the natural order of things.

The Ancient Idea of the Masses and the Elites

Originating in the Greek term for dough, *maza*, the term 'mass' signifies a shapeless, compact substance, composed of many seemingly similar units that cannot be distinguished from each other. A mass is an object ready to be shaped by somebody.[1]

When used metaphorically, as in the terminology of the masses of common men and women, the notion of the masses may be used simply as a quantitative term to signify the majority or multitude, or it may be used as a qualitative term, ascribing certain qualities to that multitude. In both cases it is invariably accompanied by its counterpart – the notion of the elites. The relationship between the masses and the elites is taken to be one of opposition and hierarchy – just waiting to be turned upside down – between a large group of subjects and a smaller group of masters. Each group is characterized by homogenous features. The elites occupy power positions in the economic, political and intellectual systems; the masses do not. Although often used in political contexts, the concept of the masses is more easily understood as a social concept in the first place: it presupposes the position of an outside observer to catch sight of the masses who cannot be seen from within by political participants.

Both notions have been significant in modern, Western social thought,[2] but are in fact neither particularly modern nor particularly Western. They can be seen as pre-modern exemplars of social categories or groups.

The idea that members of a society are divided into the masses and the elites has been influential also in pre-modern times[3] and in non-Western cultures. Thus, the assumed dichotomy of the masses versus the elites has been influential not only in the histories of mainly Christian cultures but also in Islam[4] – and in social science. It is incompatible, however, with understandings of citizenship that use the classical Aristotelian notion of the citizen as their point of departure.

Citizens, in the latter interpretation, were defined by political equality with other citizens. They were co-responsible peers, had no masters among them and no subjects. Indeed, politics was defined by being liberated from the masters versus subjects relationships that were abundant in households. All citizens were supposed to have an equal say in public matters and to carry their share of political and administrative obligations. Together, they constituted the *koinonia politike*, in Greek, or the *societas civilis*, in Latin – a civil society.[5] But citizenship was not for all.

While gradually extending citizenship to include, in principle, all adult inhabitants of a state, modern democracies have maintained an affinity for the ideal of citizens as political equals. At the same time, the extension of citizenship has been accompanied by long-standing habits of political, social and economic inequality dating back to medieval understandings of the hierarchical order of societies. In practice, ingrained social prejudices and status schemes have affected modern understandings of citizenship and the citizenry or public.

Different Western cultures have proceeded along somewhat different lines,[6] but all have been influenced by the fact that, increasingly, modern societies have acquired the features of economies. Gradually, thus, the logic of the household – *oikos* in Greek, the root of 'economy' as a term – has become dominant. In classical political thought, the household was seen as the very seat of inequality and the hierarchical exercise of power. The idea(l) of citizens as political equals constituted a countermeasure to the logic of the households. In modern societies – or economies – assumptions about masters versus subjects relationships have become staples of political thought, and understandings of the public as the people 'in contrast with those who govern them' have become commonplace.[7] Even in political life, social concepts, referring to status relations, have come to prevail.

In English, the notion of citizen – defined in the *General English Dictionary* from 1740 as a 'freeman or inhabitant of a city'[8] – gradually lost out to the notion of commoners, defined by their (lack of) financial capacities.[9] During the nineteenth century, then, the notion of the middle classes gained momentum as a term for those commoners who did not belong to the working classes but were ascribed a capacity for social ascent if sufficiently ambitious. Connected to ideals of industry and education, the terminology of the middle classes formed part of a conceptual cluster that also included commitment to modern science, to progress and to manufacturing and trade.[10]

Social concepts such as the middle classes, share their focus on status relations with the notions of the masses and the elites and introduce a tension with the idea(l) of citizens as political equals.[11] That tension, in turn, is crucial to reflections on the potential audiences of or participants in science communication. Who should be addressed – the public perceived as a mass-audience or the public perceived as co-responsible citizens?

The idea of the masses, it should be kept in mind, is based on assumptions about their ignorance and lack of intellectual capacity and, thus, their supposedly severely restricted ability to understand complicated issues. Such assumptions, of course, hamper public exchanges on science-related political issues as well as political communication in more general terms. Used as the point of departure for appeals to the general public they may become self-fulfilling and call forth precisely those qualities they are addressing. Because of their potentially far-reaching consequences, next I look into the assumptions and their backgrounds in some detail.

Of particular topicality to science is the fact that it has never been obvious where to place science and scientists. Should science be connected to the masses or to the elites? Historically, modern science has been tied to democracy, to popular rule. The tendency, however, to discursively separate scientists from the public or the citizenry at large indicates that scientists are perceived as an elite group. Individual scientists and groups of scientists identify differently. As an institution, science has never really made up its mind. There is ambiguity and tension. Considering the significance – *to* science communication and *in* the social sciences – of the view that society is composed of the masses and the elites confronting each other, the ambiguity gives food for thought. Why is it that science is linked, at the same time, both to the masses and the elites?

There might not be any sensible answer to the question of whether science rightly belongs with the masses or the elites. Maybe it is simply not a sensible question to ask. Maybe it does not make sense at all to tie science, as a body of knowledge and rational methodology and as an intellectual enterprise, to social categories. Maybe these kinds of connections serve merely to reinforce social prejudices and to lead exchanges about science-related issues astray, diverting attention from the substance of issues. It is no law of nature that scientific and other intellectual activities, such as science communication, must be perceived as expressions of social relations. It is perfectly possible to understand them simply as intellectual activities of enquiry. As it is, however, the dominant models of thought on science communication are heavily influenced by the idea that society is composed by the social categories of the masses and the elites.

The modern inversion of the ancient idea

There is, as noted, nothing particularly modern or Western about the view that society is composed of masses and elites, but an evaluative change gained momentum in the West from the late eighteenth century onwards. Closely tied to what has aptly been termed 'the invention of the people'[12] a positive valuation of the supposed masses was substituted for the hitherto negative valuation. A normative inversion occurred, maintaining widespread assumptions about the qualities of the masses, but normatively standing them on their heads. The terminology of the people and the common man converted the supposed masses into an object of worship rather than of contempt. The basic assumptions, however, were left unchanged.

Of specific relevance to our issue is the fact that assumptions about widespread ignorance and lack of intellectual capacity in the supposed masses were upheld but romanticized or sentimentalized. An intuitive wisdom – or a capacity for gut feeling, originating in inherent moral qualities – was ascribed to the masses of the people. Wit, on the other hand, had for some time been connected to academic schoolmen and other repressive elites and had experienced a decline as a term of praise.[13]

Critical enquiries into the notion of the masses as a qualitative term have connected it to a kind of person – that has come to be, or to be perceived to be, common – who is motivated primarily by the immediate prospects of pain, pleasure and gain; who is caught up in concerns with his or her private affairs; and who is highly emotional, easily manipulated and disinclined to engage in any kind of abstract thinking.[14] Positive valuations of such assumed qualities have been using instead a vocabulary that emphasizes warm-heartedness, the ability to be present here and now (as opposed to the past and the future and to faraway places) and a capacity for close relations and community building.

Core features of the idea of the masses – lack of power, personal distinction and intellectual inclinations – in short, remained constant, but were increasingly seen as positive and virtuous rather than negative, by liberal and socialist thinkers alike.

A positive valuation of the masses became manifest during and in the wake of the American War of Independence and has been connected to a wave of fascination with quantitative knowledge – an early data craze, if you like. According to historian Gordon S. Wood, it became fashionable to establish collections of facts, and

[p]eople now [in the early nineteenth century] described society more and more as a 'mass' and for the first time began using this term in reference to 'almost innumerable wills' in a positive, nonpejorative sense. The individual was weak and blind, said George Bancroft[15] in a common reckoning, but the mass of people was strong and wise. From all this followed, too, a new appreciation of statistics: in 1803 the word 'statisticks' first appeared in American dictionaries.[16]

Cutting the Earthly Chains

Combining drama and meta-religious connotations, visions of cutting the earthly chains, of liberation from limitations and uncertainties relating to space and time, body and mortality are typical themes of literary science fiction, of missionary science communication and sometimes of PR exercises by scientific institutions. One of the reasons, for instance, for NASA's financial support of the 2015 science fiction film *The Martian* was, it has been noted, that the film would make 'NASA look awesome, and a mission to Mars real'. The themes include space travel and human colonization of outer space as well as the production of humanoid robots and new human or post-human master races with strongly increased capabilities.

Such visions, some of them older than modern science, have been one of its companions from its early days. And at least one of them has suffered from its realization by way of scientific and technological development. For ages, *Ars volare*, the art of flying, was envisioned in narratives, and there appears to have been no end to the awe caused by the ascent of the first manned balloon in Paris in November 1783. A contemporary report described it as 'the most astounding achievement the science of physics has yet given to the world' and observed that the crowd gathered to follow the experiment was composed of '[t]wo hundred thousand men, lifting their hands in wonder, admiring, glad, astonished; some in tears for fear the intrepid physicists should come to harm, some on their knees overcome with emotion, but all following the aeronauts in spirit'.

When aeroplanes were actually developed and put into extensive use, and human beings did not seem to change fundamentally, Ars volare disappeared from the repertoire of visions connected to the ultimate liberation of humankind by science. Roughly since the 1950s, space travel – recently supplemented by virtual reality – has taken the place formerly occupied by the art of flying.

The textual snapshot about cutting the earthly chains is a modified excerpt from Gitte Meyer, 'Fascinating! Popular Science Communication and Literary Science Fiction: The Shared Features of Awe and Fascination and Their Significance to Ideas of Science Fictions as Vehicles for Critical Debate about Scientific Enterprises and Their Ethical Implications'. It refers to Ryan Bradley, 'Why NASA Helped Ridley Scott Create "The Martian" Film' for the statement about NASA's support for a science fiction movie and to Martin Schwonke, *Vom Staatsroman zur Science Fiction: Eine Untersuchung über Geschichte und Funktion der naturwissenschaftlich-technischen Utopie* for the notion of *ars volare*. The descriptions of the first manned balloon flight in Paris were quoted by Robert Tavernor, *Smoot's Ear: The Measure of Humanity*, 177, 116.

On the assumption that quantitative knowledge – like the marketplace – was accessible to all, such knowledge was increasingly linked to democracy and the positive idea of the common man.[17] Gradually, the popular and the commercial became almost synonymous terms, while the intellectual – as connected to learning and leisure – came to be seen as the elitist opposite of the popular.[18]

The view that society is divided into the masses and the elites has remained a premise of social thought and continues to give rise to conflicting interpretations and valuations. The notion of the masses – and, thereby, the assumed dichotomy that it forms part of – is a contested concept[19] and even those who adopt it as a model for thought disagree on its possible connections to, for instance, the notions of mobs and crowds and the concept of civilization.

Leisure, learning and social distinction

An acute awareness of their own frailty is characteristic of modern civilizations. The fear of barbarism is never far away. At the same time, there is disagreement on the very definition of civilization and barbarism, respectively, and how they may be linked to the masses or to the idea of the masses. To some, civilization is an outcome of a mass society. To others, barbarism is an outcome of a mass society. Both understandings tend to be rather intimately tied to understandings of science, viewed, in the most extreme versions, as the highest form of civilization or as an expression of modern barbarism.

French sociologist Gustave le Bon (1841–1931) carried out an early attempt to explore mass societies in a scholarly fashion. He published *The Crowd: A Study of the Popular Mind* in 1896. Since then, many have followed in his wake, but the overall understandings of and approaches to mass communication that he espoused have remained remarkably stable and now seem to constitute a tradition of modernity.[20]

Le Bon, like many others, was fascinated by the phenomenon of crowds and, apparently, took them to be an expression of strong natural forces. Also like many others, he evidently found it difficult to distinguish between masses, crowds and, for that matter, mobs. He appears to have been using the notions intermittently and did not even distinguish between a crowd gathering spontaneously in the street and deliberative assemblies such as parliaments or juries. To all such groups he attributed an 'extreme mental inferiority', connected, as he saw it, to the unconscious – 'the genius of crowds'– as opposed to the faculty of reasoning.[21]

In the introduction, he connected 'the era of crowds' to 'the creation of entirely new conditions of existence and thought as the result of modern scientific and industrial discoveries'. He also disclosed his position concerning the relationships between crowds, civilization and barbarism. Although operating with the possibility that crowds might be 'virtuous and heroic' it was his general assumption that '[c]ivilisations as yet have only been created and directed by a small intellectual aristocracy, never by crowds. Crowds are only powerful for destruction. Their rule is always tantamount to a barbarian phase.' In crowds 'the foolish, ignorant, and envious persons are freed from the sense of their insignificance and powerlessness, and are possessed instead by the notion of brutal and temporary but immense strength'.[22]

Crowds, according to le Bon, only expressed 'those mediocre qualities which are the birthright of every average individual. In crowds it is stupidity and not mother-wit that is accumulated'. As units of a crowd, individuals were easily impressed by words and images; crowds were 'credulous and readily influenced by suggestion'; they showed 'servility in the face of a strong authority'; and they were 'extremely conservative' and 'hostile to changes and progress'. Thus, it was 'fortunate for the progress of civilisation that the power of crowds only began to exist when the great discoveries of science and industry had already been effected'.[23]

Le Bon managed to crowd into a few sentences all those supposed markers of masses that have since been repeated over and over again in writings on the masses: 'It will be remarked that among the special characteristics of crowds there are several – such as impulsiveness, irritability, incapacity to reason, the absence of judgment and the critical spirit, the exaggeration of the sentiments, and others beside – which are almost always observed in beings belonging to inferior forms of evolution – in women, savages, and children, for instance.'[24]

Due to the fact that crowds were 'far more under the influence of the spinal cord than of the brain' even intelligent persons turned stupid in a crowd, le Bon found. Accordingly, '[f]rom the moment that they form part of a crowd the learned man and the ignoramus are equally incapable of observation'. On this point, he did make an exception though. Men of learning, he meant, only assumed 'all the characteristics of crowds with regard to matters outside their speciality'.[25] Apparently he expected specialized knowledge to provide its bearers with some sort of immunity.

Of particular interest to the topic of science communication are le Bon's guidelines for addressing a crowd: 'An orator wishing to move a crowd must make an abusive use of violent affirmations. To exaggerate, to affirm, to resort to repetitions, and never to attempt to prove anything by reasoning are methods of argument well known to speakers at public meetings', he found, adding: 'The art of appealing to crowds is no doubt of an inferior order, but it demands quite special aptitudes.'[26]

Speakers would have to appreciate that crowds were 'powerless [...] to hold any opinions other than those which are imposed upon them', were rather indifferent to everything that did not 'plainly touch their immediate interests', and that a 'chain of logical argumentation' would be 'totally incomprehensible to crowds'.[27]

To exercise any influence, therefore, ideas suggested to crowds should be presented in a 'very absolute, uncompromising and simple shape'. Far-reaching modifications were required in particular when 'somewhat lofty philosophic or scientific ideas' were presented. They had to be lowered 'to the level of the intelligence of crowds'. Sadly, those modifications – consisting among other things in the presentation of ideas as sentiments – always tended to be 'belittling and in the direction of simplification', but it could not be helped. Crowds were, as far as ideas were concerned, 'always several generations behind learned men and philosophers'. Come to that, most people – here, le Bon actually referred to most people, rather than to crowds – were unable to shape an opinion of their own by way of reasoning and did not understand statistics.[28]

An orator in 'intimate communication with a crowd' could, le Bon observed, 'evoke images by which it will be seduced'. Discussion with crowds, on the other hand, was out of the question. Judgements accepted by crowds were 'merely judgments forced upon them and never judgments adopted after discussion'.[29]

Fear of the barbarians: Variations on a theme

A contemporary of le Bon, American sociologist Thorstein Veblen (1857–1929) was concerned with related topics. His *The Theory of the Leisure Class* was published in 1899 and is another example of the modern fear of barbarism. Veblen's ideas of barbarism, however, did not correspond to le Bon's ideas. Both shared the view that society was divided into the masses and the elites, and both presented themselves as friends of science and progress. Otherwise, their valuations were different. Veblen, as opposed to le Bon, identified with the assumed masses, with 'the vulgar' as opposed to 'their masters'.[30] To him, the elites were a model of barbarism and conservatism, and his identification with science was combined with a fierce aversion to the humanities.

The elites, in Veblen's terminology, constituted a 'leisure class'. And Veblen held leisure in contempt. Leisure signifies free time. In antiquity, such free time was much appreciated and came with connotations of dignified activity. In fact, we continue to refer to free time in that sense every time we mention a school – the term originates in a Greek term for free time.[31] To Veblen, however, with his background in Puritanism, the fact that leisure connoted non-productive consumption of time was offensive or downright scandalous.

Leisure, he found, was 'closely allied in kind with the life of exploit'. He linked it to 'the knowledge of dead languages and the occult sciences; of correct spelling; of syntax and prosody; of the various forms of domestic music and other household art; of the latest properties of dress, furniture, and equipage; of games, sports, and fancy-bred animals, such as dogs and race-horses'. All this and much more – including refined tastes, manners and habits of life – was confined to the habitats of the leisure class 'because good breeding requires time, application and expense, and can therefore not be compassed by those whose time and energy are taken up with work'. In contrast, 'productive labor' was the hallmark of 'the working class' or 'the lower classes' or 'the mass' or 'the people'.[32]

In 'barbarian culture', Veblen noted, the upper classes were exempt from industrial employments and the leisure class constituted a 'superior pecuniary class', characterized by 'the requirement of abstention from productive work', which it took to be 'a mark of inferiority'. Therefore 'vulgarly productive occupations' were 'unhesitatingly condemned and avoided' and taken to be 'incompatible with life on a satisfactory spiritual plane – with "high thinking"'. Government and war, according to Veblen, served as the main sites of occupation of the higher leisure class. Both occupations were 'of the nature of predatory, not of productive, employment' just as politics and law were useless activities.[33]

An enthusiastic supporter of industrialism, Veblen described the 'industrial virtues' as 'peaceable traits' marked by 'an impersonal, non-invidious interest in the work at hand'. Those virtues, he assumed, were widely distributed 'among the classes given to mechanical industry' and to 'the collective life'. The masses, he assessed, were primitive, but partaking in (rational) industrial activity, they were likely to become increasingly rational themselves. Thus, insofar as the industrial virtues became dominant, humankind could, Veblen was convinced, look forward to a bright future of peace and prosperity. The 'inertness of the mass of any modern civilized community' made war a highly improbable prospect. And habits of using impersonal reasoning to identify, from one case to another, a 'quantitative causal sequence' would serve to further efficiency and effectivity. All in all, Veblen found, the 'habit of mind which best lends itself to the purposes of a peaceable, industrial community, is that matter-of-fact temper which recognizes the value of material facts simply as opaque items in the mechanical sequence'.[34]

Conversely, the leisure class acted to 'lower the industrial efficiency of the community and retard the adaptation of human nature to the exigencies of modern industrial life'. Conservatism, Veblen argued, was an upper-class characteristic whereas innovation was a lower-class phenomenon, deemed vulgar by the leisured and 'parasitic' upper classes. Their appreciation of handmade objects was mere leisure-class snobbery and their 'veneration for the archaic or obsolete, which in one of its developments is called classicism' was simply a 'secondary expression of the predatory temperament'. Veblen did not really find a place for the middle classes in his templates, but simply did away with them in a side remark about 'the lower or doubtful leisure class in America – the middle class commonly so called'.[35]

The concluding chapter concerned '[t]he Higher Learning as an Expression of the Pecuniary Culture'. According to Veblen, it was 'in the higher learning, that the influence of leisure-class ideals is most patent'. The 'recondite element in learning', he argued, was still, 'as it has been in all ages, a very attractive and effective element for the purpose of impressing, or even imposing upon, the unlearned; and the standing of the savant in the mind of the altogether unlettered in great measure rated in terms of intimacy with the occult forces'. The activities called higher learning – and in particular those schools whose chief end was 'the cultivation of the "humanities"' – lacked, according to Veblen, any positive significance for the life of production. Thus, it came as no surprise, that the 'ritualistic features of the educational system' had their place 'primarily in the higher, liberal and classic institutions and grades of learning, rather than in the lower, technological, or practical grades, and branches of the system'.[36]

Veblen connected 'the truly learned' to 'that field of learning within which the cognitive or intellectual interest is dominant – the sciences properly so called'. Turning to science 'in the sense of an articulate recognition of causal sequence in phenomena' he made the case that 'while the higher learning in its best development, as the perfect flower of scholasticism and classicism, was a by-product of the priestly office and the life of leisure, so modern science may be said to be a by-product of the industrial process'.[37]

The future belonged to science as opposed to the humanities, and to the masses of the people as opposed to the leisured elites. 'The sciences have been intruded into the scholar's discipline from without, not to say from below,' Veblen noted and expanded:

> It is noticeable that the humanities which have so reluctantly yielded ground to the sciences are pretty uniformly adapted to shape the character of the student in accordance with a traditional self-centred scheme of consumption; a scheme of contemplation and enjoyment of the true, the beautiful, and the good, according to a conventional standard of propriety and excellence, the salient feature of which is leisure – otium cum dignate.[38]

That, however, was completely out of touch with 'the everyday life and the knowledge and aspirations of commonplace humanity'. To the aim of realizing 'an efficient collective life under modern industrial circumstances', classical learning was worse than useless.[39]

Interestingly, thus, Veblen and le Bon were fundamentally in agreement with respect to significant aspects of how to address the supposed masses of the people. Identifying with opposite sides of an assumed societal dichotomy, taken for granted by both, they shared basic assumptions about dominant features of the masses. According to those assumptions, intellectual appeals to the masses, going beyond matter-of-fact statements, would be counterproductive.

The modern reinvention of the laity

Thorstein Veblen and Gustave le Bon both identified with progress, took themselves to be promoters of an idea(l) of science – not necessarily the same idea(l), though – and declared their opposition to conservatism. But, whereas Veblen identified with the supposed masses of the people, le Bon identified with the supposed elites.

In current usage, the social categories of the scientific experts and the ordinary citizens have come to be widely used as synonyms for (intellectual) elites and (lay) masses. The concept of the laity – inherited by science from the medieval church – has been smoothly fused with the concept of the masses.

Literally, the concept of the layperson signifies a person who is in lack of knowledge, and precisely that quality – or lack of quality – has continuously been attributed to the masses. Probably from a very early stage, the use of the concept of the laity, originally signifying a lack of religious knowledge, has also implied other connotations, informed by a rich legacy of social prejudice. Such expressions as the 'meaner sort of people', 'the common and meaner sort', 'the lower orders' and 'the rabble'[40] obviously were staples of seventeenth-century discourse. They were related to knowledge and learning in such combinations as 'the unknowing multitude'[41] and 'the rabble that cannot read'.[42]

Gradually, openly abusive characterizations of the masses have been replaced by less immediately demeaning labels. Typically, when, in 1929, science fiction writer H. G. Wells (1866–1946), his son, zoologist G. P. Wells (1901–1988) and biologist Julian Huxley (1887–1975) launched their popular science magazine, *The Science of Life*, they declared the publication to be targeting 'the ordinary man'. They also repeatedly emphasized their belief in the superiority of science and their contempt of the crowd. Denouncing '[v]ulgar fashions, false interpretations and decaying traditions' they noted: '[T]he crowd is always about us; but we forget that these things are divergent and inconsecutive and accumulate no force, while scientific work and lucid thought are persistent and cumulative.'[43]

At that time, the notion of science popularization had been in use for about a century.[44] Although nowadays sometimes accompanied by slight misgivings,[45] it is still widely used as a general science communication term and appears, at first glance, to be a direct offspring of the concept of the public – adult

Standardization for the Masses

The development of tinned foods relied on modern science. The successful processes behind tinned food depend on knowledge from quite a few scientific fields. And there are other significant interconnections between tinned food, in particular, and modern, science-based technology, in general. Tinned food is standardized and homogenized. One can is precisely like the other. Moreover, the technological aim of making life easier is clearly realized by the production of tinned food. No wonder, then, that when tinned food became widely available in the early twentieth century, in heated debates it came to serve as a stand-in for modern science and modernity in general.

In line with 'false teeth and other modern nonsense', tinned food was characterized as 'mechanical and soulless', a 'homogenized mass-product'. Thus, the cans were even used to emphasize a seeming link between a perceived mass public and modern science, preparing the way for a view of this particular method of food preservation as a kind of democratization and, even, for the view that a preference for fresh food was somehow elitist.

Food preservation and production have moved on since then, supported by scientific and technological developments. To some extent, fast food has replaced tinned food as a symbol of all things mechanical and soulless. Today, however, science appears on both sides of the argument. Opponents of fast food have access to a huge cache of scientific data that may be used as evidence – altering the focus – that fast food is a threat to public health. As always, however, when attempts are made to use science to resolve substantial disagreement there will be a supply of scientific counter-evidence. As the argument is turned into a methodological battle about how best to measure and organize the most effective way of feeding the human organism, the aversion to the mechanical and soulless will have to look for yet another outlet.

The textual snapshot about standardization for the masses has drawn on John Carey, *The Intellectuals and the Masses: Pride and Prejudice among the Literary Intelligensia 1880–1939*, 21–22.

population, from *populus*[46] – which in modern democratic societies constitutes the citizenry. Popularization of science, however, presupposes the existence of popularizing scientists who, although adult, apparently do not form part of the public. It seems a reasonable interpretation that the idea of popularization should be connected not to the *political* concept of the public as citizenry but rather to the *social* category of commoners or plebeians, originating in a Greek term for crowd or throng and, thus, a member of the family of concepts used to denote the masses as opposed to the elites.[47]

By adopting the concept of the laity, now appearing in the hybrid shape of lay masses, science as an institution established a hierarchical order, informing understandings of science communication in which scientists constituted an elite group of knowers. Even though many individual scientists might identify otherwise, mainstream understandings of science communication did not allow them to escape an elite identity, in one shape or another.

Education and eugenics

To some, evidently, it has neither demanded a lot of effort to think of society in terms of the elites and the masses nor to achieve an elite identity of assumed superiority to the masses, complete with an assumed obligation to improve the quality of those masses by subjecting them to various kinds of therapy. All along, education in the shape of science popularization has been widely taken to constitute one possible form of therapy. Time and time again, eugenics has been advocated as another such possibility.

Among numerous possible examples,[48] the popular science magazine published by Wells, Wells and Huxley between 1929 and 1930 again presents itself as particularly compelling. In line with Thomas Sprat's twin aims of fighting the foes of ignorance and false opinions – proclaimed almost three centuries previously in his early history of the Royal Society – the magazine seems to have been launched to educate the general public *and* to further the cause of eugenics: 'For a number of generations, at any rate', according to the magazine,

> a dead-weight of the dull, silly, under-developed, weak and aimless will have to be carried by the guiding wills and intelligences of mankind. There seems to be no way of getting rid of them. The panics and preferences of these relatively uneducable minds, their fat and foolish tastes, their perversities and compensatory loyalties, their dull, gregarious resistances to comprehensive efforts, their outbreaks of resentment at any too lucid revelation of their inferiority, will be a drag, and perhaps a very heavy drag, on the adaptation of institutions to modern

needs and to the development of common knowledge and a common conception of purpose throughout mankind.[49]

Thanks to the progress of science, however, that sad state of human affairs would not last forever. The authors looked forward to 'the dominance of a collective control of human destinies', envisioning 'a time when the species will have a definite reproductive policy, and will be working directly for the emergence and selection of certain recessives and the elimination of this or that dominant'.[50]

A good many other prominent scientists and science enthusiasts argued along related lines. In the 1920s, concerned by the perceived irrationality of the general public, American political scientist Charles Merriam (1874–1953) – one of the fathers of behaviourism – was an advocate of 'civic education' with the primary goal of inculcating the scientific method. In 1925, he added eugenics to his programme, pointing to 'two great mechanisms' that he expected to be equally effective: 'education and eugenics'.[51]

At about the same time, similar arguments were made by, for instance, American geneticist Hermann Muller (1890–1967),[52] who was awarded the Nobel Prize in 1946. Both Huxley and Muller were still active and sticking to the promotion of eugenics in the early 1960s.[53] In 1962, Muller – addressing a group of prominent fellow biologists – made the case that the quality of the public had to be improved to safeguard democracy: 'Unless the average man can understand and appreciate the world that scientists have discovered', he would, Muller feared 'fall into the position of an ever less important cog in a vast machine [...] Democratic control, therefore, implies an upgrading of the people in general in both their intellectual and social faculties, together with a maintenance or, preferably, an improvement in their bodily condition'.[54]

In the ensuing debate, Francis Crick (1916–2004) argued that much improvement might be achieved by 'simply taking the people with the qualities we like, and letting them have more children'. Concerns about 'the risk that biology will not be taught objectively' were briskly whisked away by Julian Huxley: '[W]e must let the biological profession itself do the job.' During the discussion, interconnections among 'creativity, intelligence, and the leaning towards science' seem to have been taken for granted,[55] and 'man's biological future' was envisioned as 'his future as a scientist'.[56] Science, apparently, was not only entitled but also duty-bound to recreate humankind, using an image of the scientist as the model.

Shuttling between Elitism and Populism

As a qualitative notion, I have been arguing, the term 'the masses' ascribes certain features to the majority of the population, the multitude. These features include emotionality, as opposed to intellectual leanings, and a propensity to concentrate on the here-and-now as opposed to all that is farther away in either space or time. From an elitist point of view – top down – such assumed features give cause for contempt. Maintaining the basic assumptions but reversing the normative assessment, we arrive at a positive understanding of the supposed masses in the shape of the people as a warm-hearted collective of ordinary men and women, all motivated primarily by everyday concerns, unlettered, dedicated to local and community affairs, driven by deeply rooted moral instincts and, importantly, uncontaminated by vested interests, ascribed only to the elites.[57] That valuation, in turn, prepares the way for populism as inverted elitism, celebrating instead of expressing contempt for the unknowing multitude.

Because the basic assumptions remain unchanged, populism may easily revert to elitism, and vice versa, creating an elitism–populism axis.[58] Nowhere along that axis are intellectual capacities ascribed to the general public. The idea of an intellectual deficit in the general public is one of its founding features. Against that background, it seems relevant to ask to what extent understandings of science communication are and have been linked to that kind of axis, deriving in a non-egalitarian context and likely to reproduce that feature over and over again.

Interestingly, both Thorstein Veblen, representing typical populist views, and Spanish philosopher, writer and politician José Ortega y Gasset (1883–1955), representing typical elitist views, connected science, viewed as a purely technical enterprise, to the masses. And returning to Thomas Sprat, we have found that he, in the mid-seventeenth century, exhibited a preference for 'the language of Artizans, Countrymen and Merchants, before that, of Wits, or Scholars'.[59] Long before populism was coined as a (positive) term in the United States in the late nineteenth century,[60] populist attitudes may have been informing the idea of science as a cause, closely tied to the cause of the people. But how does that combine with the equally old idea of a lay and deficient public, inferior to scientists, indicating rather an elitist understanding of science?

The ambiguities make sense if understandings of science communication have actually been tied continuously to an elitism–populism axis, widely assumed to form part of the natural order of things and only allowing movement between its poles. That interpretation might even help us understand why exchanges on science communication have been continuously pervaded by the idea of just two groups, scientific experts and the lay masses of commoners, whether sharply contrasted in top-down approaches or urged to enter into dialogues.

At some times and in some places, populist valuations gain momentum, only to be superseded at some stage by their elitist counterparts. The pendulum swings back and forth and back again. This was particularly obvious during the Progressive Era of the late nineteenth and early twentieth century in the United States.

The divergent arguments advanced by, respectively, American philosopher John Dewey (1859–1952) and American journalist and commentator Walter Lippmann (1889–1974) in the 1920s can be seen as a model of conflicts – that have, since then, been repeated over and over – between populist and elitist valuations of the masses. Dewey presented his argument in *The Public and Its Problems*, published in 1927,[61] and Lippmann made his case in *Public Opinion*, published in 1922.[62]

Both writers shared a deeply rooted belief in science, a view of humans as social animals and an understanding of the public as the masses of the people, as opposed to the elites. Moreover, both apparently subscribed to a range of assumed dichotomies, counterposing for instance the individual and society. They took different sides, however.

Lippmann saw humans as selfish and narrow-minded social animals who were equally unable and disinclined to think beyond their own immediate interests. As individuals he took them to be at war with society. Dewey's social animals, on the other hand, were fundamentally kind-hearted and marked by intuitive moral instincts and a longing for consensus, unity, commonality and intimacy, corresponding to his vision that society might evolve into a 'Great Community'. The local, according to Dewey, 'is the ultimate universal, and as near an absolute as exists'.[63]

Dewey wanted science 'absorbed and distributed'. He also advocated the further expansion of science. In a 1946 afterword to a re-publication of his book, he advanced the expectation that the use of scientific methods would promote 'effective foresight of the consequences of social policies and institutional arrangement'. He furthermore suggested that 'a considerable part of the remediable evils of present life are due to the state of imbalance of scientific method with respect to its application to physical facts on one side and to specifically human facts on the other side'. The most 'direct and effective way out of these evils', Dewey found, was 'steady and systematic effort to develop that effective intelligence named scientific method in the case of human transactions'.[64]

Lippmann was an equally firm believer in social science as a political tool but did not share Dewey's optimistic view of humans as social animals and was far from impressed by the general human capacity for reason. 'The mass of absolutely illiterate, of feeble-minded, grossly neurotic, undernourished and frustrated individuals' was, Lippmann wrote, 'very considerable'.[65]

It was the overall argument of his introductory chapter –'The World Outside and the Pictures in Our Heads' – that the human senses were untrustworthy. Humans could not avoid interpreting what they saw, but interpretative activity was erratic by definition. Interpretations constituted a sort of illusion. Therefore, public opinion was founded on mere fictions. Social science was a much more reliable guide to comprehending reality, he argued: 'In the absence of institutions and education by which the environment is so successfully reported that the realities of public life stand out sharply against self-centred opinion, the common interests very largely elude public opinion entirely, and can be managed only by a specialized class whose personal interests reach beyond the locality.'[66]

Lippmann did not see it as a purpose 'to burden every citizen with expert opinions on all questions, but to push that burden away from him towards the responsible administrator'. He and Dewey shared the assumption that science was above power relations and partisanship – in Lippmann's formulation the 'value of expert mediation' was not 'that it sets up opinion to coerce the partisans, but that it disintegrates partisanship'[67] – and both were suspicious of power. They disagreed, however, on the relationship between the people and power.

In contrast to Lippmann's assessments, Dewey apparently placed the people on the same side as science, in an imagined sphere beyond the exercise of power and power relations. He was in line with other 'Progressives', not least from the fields of social science and journalism who, it has been argued, 'were all distinguished by their emphasis on facts over general interpretation. In each a passionate spirit of advocacy was submerged in a confident expectation that the public could learn, and draw appropriate conclusions, from a scientific treatment of the facts.'[68]

In 1922, in a review published in the *New Republic* – edited by Lippmann – Dewey described Lippmann's *Public Opinion* as 'perhaps the most effective indictment of democracy as currently conceived ever penned'.[69] Five years later, then, Dewey published his defence of democracy.

While Dewey's reasoning pointed in the direction of populism, Lippmann rather argued along technocratic – and thus elitist – lines. Along with many others, they shared the assumption, it appears, that those were the options and, thus, that an elitism–populism axis formed part of the natural order of things.[70]

Ambiguity: Science, the masses and the elites

The Dewey–Lippmann controversy epitomizes a tension that seems to have been present in modern science from its very beginnings. While the view of the general public as the masses has been a social science feature all along – dependent, as that view is, on the position of an outside observer – it has never been clear-cut where science rightly belonged. Should science be placed on the side of the masses or on the side of the elites? Was science common sense or much superior to common sense? Was science an intellectual endeavour and, if so, in what sense – and would that turn scientists into an intellectual elite, radically separated from and perhaps even in opposition to the general public?

Different understandings of the concept of knowledge constitute a common denominator of all those and many related conflicts. The identification, cultivated by Veblen, with a purely technical concept of knowledge as specialized no-nonsense know-how appears, more often than not, to be tied to positive valuations of the masses with their supposed lack of wider intellectual capacities and inclinations. Those interpretations place science on the side of the masses, perceived as their means of guidance. Conversely, wider understandings of knowledge, separating science less rigidly from humanist learning, have been frequently connected to elitist contempt of, at the same time, the masses and science in the above sense. Thus, generally they have been equally unable to escape the elitism–populism axis.

Actually, Ortega y Gasset in his *The Revolt of the Masses*, published in 1930, attempted to introduce a, to some extent, different twist. He placed the scientist as technician in the masses only to make the case that science, as a pillar of modern civilization, did not rightly belong there.

The 'common' or 'average' man, according to Ortega y Gasset, had 'learned to use much of the machinery of civilisation, but […] is characterised by root-ignorance of the very principles of that civilisation'. Subscribing to mainstream understandings of the masses, he saw 'mass-man' as 'a primitive' who had 'no attention to spare for reasoning', who learned 'only in his own flesh' and had 'no interest in the basic cultural values'. Like members of the hereditary aristocracy of former times, mass-men had the qualities of spoilt children and, sadly, he found, the 'prototype' of that sort of human being was now – the scientist. Science itself, 'the root of our civilisation', Ortega y Gasset claimed, 'automatically converts him into mass-man, makes of him a primitive, a modern barbarian', finds 'a place for the intellectually commonplace man and allows him to work therein with success'.[71]

Science, as Ortega y Gasset saw it, had been perverted by increasing specialization that had undermined its intellectual qualities. Over a few generations the enterprise that set out with the writing of encyclopaedias[72] had been reduced to narrow-minded specialization. The majority of scientists, so Ortega y Gasset argued, helped 'the general advance of science while shut up in the narrow cell of their laboratory, like the bee in the cell of its hive', knowing their 'own tiny corner of the universe' but being 'radically ignorant of all the rest'.[73]

Snapshot VII

Hype, Secrecy, Xenotransplantations

Transplantation of organs from animals to humans has been envisioned, bobbing to the surface now and again, since the seventeenth century. Experimental activity was intensified in the 1960s, and in the 1990s it was linked to the biotech industry. In 1995, two British scientists caused a minor sensation when they announced their intent to carry out trials with transgenic pigs that had been altered to make their organs compatible with the human immune system. Public committees were formed to assess the prospects. Novartis, the Swiss-based multinational, bought the company formed by the two scientists and things went quieter, but in 1999 Novartis announced approaching trials. Other companies in the United Kingdom and the United States had related plans. The commercial idea was to develop packages of organs and immunosuppressing medicines.

It soon became clear, however, that there was a zoonotic risk. Infectious diseases might be transferred from donor pigs to humans and result in epidemics of diseases hitherto unknown in humans. The European Council demanded a moratorium. In Canada, Sweden, the United Kingdom and the United States, the Organisation for Economic Co-operation and Development (OECD) and the World Health Organization (WHO) committees were set up to prepare safety guidelines, but the prospective producers were not very forthcoming with technical details. All wanted to be first on the market and to set the standards. The knowledge was privately owned, but the apparent challenge was a public one. Extensive reports were prepared, and extensive safety regulations – including extensive surveillance of receivers of organs and those with whom they might exchange bodily fluids – were proposed. Odd questions were looked into: Would organs from pigs grow along with a human child? Should weak-hearted donor pigs be trained in treadmills? What would a hyper-hygienic scheme for the raising of pigs look like?

The frenzy, affecting big money and public authorities alike, was dampened when it was shown that retroviruses from pigs could actually infect human cells. It all came to a temporary standstill.

The textual snapshot about xenotransplantations is based on extensive research and writings that I carried out in the late 1990s for, in particular, the then Danish Board of Technology Assessment. Gitte Meyer, 'Knald eller fald for organer fra grise' [Neck or nothing for organs from pigs] is an example.

Previously, Ortega y Gasset mused,

> men could be divided simply into the learned and the ignorant, those more or less the one, those more or less the other. But your specialist cannot be brought under either of these two categories. [...] We shall have to say that he is a learned ignoramus, which is a very serious matter, as it implies that he is a person who is ignorant, not in the fashion of the ignorant man, but with all the petulance of one who is learned in his own special line.[74]

In the long-term, Ortega y Gasset feared, a continuation of that development, with its specialization and isolation, threatened to cut off science from its intellectual roots. Thus, it almost amounted to a betrayal of science as an intellectual enterprise.

Along related lines, science communication as the mere dissemination of scientific findings and knowledge claims, supported by dramatization but circumventing complexities, critique and sceptical questioning, can be seen as a threat to science as an intellectual enterprise. Based on the potentially self-fulfilling caricature of the general public as the masses that Ortega y Gasset subscribed to, like so many others, it might even further the coming into being of that kind of public.

The mass public as an object of social-scientific enquiry

Since the end of the Second World War, American social science has acquired the status of an international social science model. Imitated by social scientists around the globe and serving as a supplier of standards in many contexts – science communication included – it has been heavily influenced by the idea of the masses and the elites.

A much quoted 1964 article, 'The Nature of Belief Systems in Mass Publics' by political scientist Philip E. Converse (1928–2014), is a typical example that also illustrates how education came to be viewed increasingly in terms of status positions and relations.

The article concerned 'the typical state of distribution of political information in societies as we find them in "nature"' and was based on the assumption that 'there can be no doubt that educated elites in general, and political elites in particular, "think about" elements involved in political belief systems with a frequency far greater than that characteristic of mass publics'. A conservative expectation would have it that 'strict logical inconsistencies (objectively definable) would be far more prevalent in a broad public'.[75]

Converse found what he was looking for. 'First', he wrote, 'the contextual grasp of "standard" political belief systems fades out very rapidly, almost before one has passed beyond the 10 percent of the American population that in the 1950s had completed standard college training'.[76]

When beliefs moved 'downwards' – from the elites to the masses – 'the objects that are central in a belief system', according to Converse, underwent systematic change.

> These objects shift from the remote, generic, and abstract to the increasingly simple, concrete, or 'close to home'. Where potential political objects are concerned, this progression tends to be from abstract, 'ideological' principled to the more obviously recognizable groupings or charismatic leaders and finally to such objects of immediate experience as family, job, and immediate associates.[77]

Those changes, Converse added, did not constitute 'a pathology limited to a thin and disoriented bottom layer of the *lumpenproletariat*; they are immediately relevant in understanding the bulk of mass political behavior'. That behaviour, in turn, was connected by Converse to the 'limited horizons', 'foreshortened time perspectives' and 'concrete thinking' that had been 'singled out as notable characteristics of the ideational world of the poorly educated'. Presenting himself in terms of '[w]e, as sophisticated observers',[78] Converse left no doubt about his own elite identity.

Describing the American 'liberal-conservative continuum' as 'a rather elegant high-order abstraction', he found that such abstractions were 'not typical conceptual tools for the "man in the street"'. Without using the terminology of the middle classes, Converse concluded that the groups in the middle of the social and educational hierarchy had 'a clear image of politics as an arena of group interests and, provided that they have been properly advised on where their own group interests lie, they are relatively likely to follow such advice'. They lacked, however, 'the contextual grasp of the system to recognize how they should respond to it without being told by elites who hold their confidence'.[79]

All in all, according to Converse, 'almost four out of ten of the population do not have a clue, what politics is about, while another half of the population understands politics to be a simple matter of group interest'. He saw this claim as a challenge to what he took to be 'the common elite assumption that all or a significant majority of the public conceptualizes the main lines of politics after the manner of the most highly educated'. In contrast, he found a more acute sense of realism among local politicians: 'Anyone familiar with practical politics has encountered the concern of the local politician that ideas communicated in political campaigns be kept simple and concrete. He knows his audience and is constantly fighting the battle against the overestimation of sophistication to which the purveyor of political ideas inevitably falls prey.'[80]

To Converse, American society had in fact 'two populations' or 'two publics', and 'the mass of less knowledgeable people' constituted the majority population. He illustrated the widespread ignorance by statements such as: 'Some American adults would not know that Africa's population is largely Negro', and '70 percent is a good estimate of the proportion of the public that does not know which party controls Congress'. It was one of his overall conclusions that '[t]he party and the affect toward it are more central within the political belief systems of the mass public than are the policy ends that the parties are designed to pursue'. The 'common citizen', he found, as distinct from 'the truly involved citizen', failed to 'develop more global points of view about politics'.[81]

Unsurprisingly, Converse's findings did not undermine the premises of his research questions. These premises, in turn, contributed to shaping his enquiries in a very direct way. He simply used very differently formulated questions depending on whether he was approaching members of the presumed masses or representatives of the presumed elites: 'As a general rule', he argued, 'questions broad enough for the mass public to understand tend to be too simple for highly sophisticated people to feel comfortable answering without elaborate qualification'. Supposed members of the mass public, for instance, were asked to consider the following statement about employment policies: 'The government in Washington ought to see to it that everybody who wants to work can find a job.' Supposed representatives of the educated elites were asked, instead, the following question: 'Do you think the federal government ought to sponsor programs such as large public works in order to maintain full employment, or do you think that problems of economic readjustments ought to be left more to private industry or state and local government?'[82] Converse, thus, appears, on the one hand to have asked for simple-minded answers from those he took to be simple-minded whereas, on the other hand, he asked for sophisticated answers from persons he expected to be sophisticated.

Converse's assumptions and approaches were mainstream. Similar understandings formed the basis of, as another example, American sociologist Seymour M. Lipset's (1922–2006) *Political Man: The Social Bases of Politics*, first published in 1959. Lipset hypothesized that social groups marked by low income, low-status employment and little or no education were attracted to intolerance and prone to view reality in terms of rigid black and white oppositions. Those groups, therefore – as opposed to more 'sophisticated' groups – constituted a possible source of extremism. Lipset even contrasted 'intellectual reflection' to 'primitive energy'.[83]

All the usual assumptions about mass publics were present in Lipset's text: lack of knowledge and verbal capacity; a high degree of suggestibility;

lack of the ability to imagine the past and the future and to understand complicated and abstract views and ideas – indeed, a general lack of imagination and capacity for logical thought. Members of the lower classes, according to Lipset, were likely to have experienced deprived childhoods and, as a consequence, tended to display only defective capacities for reasoning and to be interested merely in trivial pursuits. Constituting an anti-intellectual force they were susceptible only to appeals of a non-complicated, plain, simple nature.[84]

Generations of social and political scientists were brought up on such understandings of the public as a mass public,[85] informing also the then emerging field of mass communication studies and influencing understandings of science communication as an elite activity aimed at improving the barbarian masses of the people.

The deficit model of the public: Criticized and persistent

The Progressives of the early twentieth century felt, it has been observed, 'a sense of responsibility, even stewardship, of a democratic mass citizenry' and were marked by a 'missionary zeal of public service'. But they also struggled with their seeming separation, as intellectuals, from 'the mass of their fellow citizens'.[86] The student movements of the 1960s and 1970s struggled no less than the Progressives with their seeming separation from the mass of the people. The populist assumption that the people were somehow disconnected from social interests – that only the elites were interested parties – served, however, to legitimize social intercourse with the masses of the people. Moreover, it was easily combined with the continued adoption of the aim – often motivated by a philanthropical spirit – of educating the ignorant masses.

That aim, in turn, has remained in force across elitist and populist valuations of those assumed masses. The assumption that the public is incompetent to discuss science-related issues was almost unanimous among the interviewees when, in 2005–6, I did a series of interviews with European bioscientists. At least, according to most of the interviewees, the general public was not in a position to engage in discussions about *their* particular field of research. Instead, the public needed to be educated about it.[87]

At that time, a critical discussion of the so-called deficit model of the public had been going on for some decades among science communication scholars. The model kept – and keeps – bobbing up again. So did – and does – the criticism.[88]

The term was introduced as a critique of condescending attitudes towards the general public. In its most basic form, it simply refers to the assumption that there is inadequate knowledge of science among the general public. In a further step, such knowledge inadequacy may be linked, as a cause–effect

Golden Rice and Harsh Reality

Some scientists were hoping rather loudly when, in the early 1990s, the Golden Rice project was launched. Developing genetically engineered rice with the capacity to produce beta-carotene – making it golden like carrots – the scientists hoped to alleviate malnutrition and prevent blindness in third world countries. Vitamin A and iron deficiency are widespread in populations depending on rice as their main source of nutrition. The golden rice came with added iron and, in principle, the human body would produce vitamin A from the beta-carotene. This example of plant biotechnology – widely unpopular at the time – with a human face only needed to be disseminated to the poor to do its good work.

A harsh clash with reality awaited the publicly spirited scientists. Employed by a Swiss public research institution, they were aiming for a publicly organized realization of their vision, but outside the laboratory was a world of complexities, uncertainties, conflicting interests and disagreement. Multinational companies, struggling to have genetically modified plants accepted by the public, were quick to spot the possible PR value of the project – and to inadvertently activate opponents in environmental organizations. Intellectual property rights became an issue. Serving as a container for deeply rooted disagreements about, among other things, property rights, the debate evolved along the lines of the well-known pattern of polarization and demonization. Accusations of irrational and religiously influenced fear went one way while accusations of cynicism and greed flew in the other direction. Meanwhile, the golden visions were gradually confronted with an increasing amount of down-to-earth problems. Coordination with other attempts to alleviate vitamin A deficiency had been ignored. Local soil and climate conditions might cause cultivation difficulties. And the human physiology might not be as cooperative as originally assumed.

The rice is still golden, but currently only in the literal sense.

The textual snapshot about golden rice draws on Gitte Meyer, 'Gylden ris har lang rejse foran sig' [Golden rice has a long journey ahead]. See also: Gerry Everding, 'Genetically Modified Golden Rice Falls Short on Lifesaving Promises', and Tom Philpott, 'Whatever Happened to Golden Rice?'.

connection, to attitudes towards science, assuming inadequate knowledge of science to result in a lack of appreciation of science: 'the role of scientific knowledge in explaining people's attitudes towards science'.[89] Critics have called this 'the knowledge-attitudes model of the Royal Society: *the more you know, the more you love it*'.[90] Yet other critics have been particularly keen to expose 'the deficit model (mis)understanding of public dissent';[91] that is, attempts to explain opposition to specific technological enterprises as the outcome of inadequate knowledge.

The critique materialized as a relatively late offspring of the student movements of the 1960s and 1970s. Some were disturbed by the assumption, evident in science communication discourses, of a deficit in the receiving end. Although that kind of assumption is a premise of any didactic effort and, thus, nothing to frown upon as a classroom phenomenon, in a wider societal context it can be seen as an expression of contempt of the people, the *demos* of democracy, and the autonomous citizen. It became a target for critique partly because of its potential to disrupt the science–democracy link, close to the heart of the movements.

Moreover, by that time the assumption of a knowledge deficit had been expanded and had come into use as an accusation that might be directed even against some scientists. During the 1960s, scientific methods were applied as means to document adverse effects of science-based technologies. Along with the environmental movement, the field of environmental science evolved as a response to the widespread application of science-based technologies in the production sphere. Science and technology critique went scientific and became, at the same time, exposed to deficit accusations; that is, to the claim that the critique was based on inadequate knowledge and appreciation of – science. Thus, probably, some of those seeds were sown that would later develop into the coining of the deficit model as a critical term within an emerging field of science studies.

The critique, however, has tended to remain tied to the understanding of science communication as a process of science dissemination and consumption and, thus, has left the fundamental framework of the didactic paradigm untouched. Much effort has been devoted to deconstructing the concept of the public, perceived as a social concept signifying a homogenous group or mass of people. The notion of 'publics' – 'locally situated groups, each of which makes sense of scientific knowledge in its own way' – has been substituted for that of 'the public'.[92] Because of the basic understanding of science communication as the communication *of* scientific knowledge, those publics, in turn, might as well be called audiences or consumer groups. In the classical political sense they are not publics any more than the notion of 'scientific citizenship' – introduced alongside the terminology of publics – captures the

Well-Being Units

Applying methods from the exact sciences, thousands of academics around the globe are currently occupied with attempts to monitor human happiness and unravel the assumed mechanisms behind it. Surveys are used to measure people's general satisfaction with life, how frequently they experience various emotional states and to what extent they perceive their lives to be meaningful. Apparently, the whole enterprise, supported by the OECD and the United Nations, is directed at facilitating the controlled production of happiness, or, in the terminology of this widespread field of research and politics, 'extra units of well-being', fitting into cost-benefit analyses, bench-marking and simplifying communication schemes.

At academic conferences, researchers present outcomes of happiness and well-being research in formats that indicate a technical-scientific approach and, thus, draw on the authority of exact science. Extensive use is made of the forceful language of numbers and of expressions, forming part of an engineering terminology, such as: applications, data, determinants, dose-responses, exposure, implementations, interventions, mechanisms, predictions, prototypes and tools.

Due to its inherent normativity and ambiguity, however, the concept of happiness – and other members of a family of related concepts – has an innate quality of contestedness. It is a thick concept, descriptive and normative at the same time. As a term, its root meaning connects to luck, fortune, coincidence – uncontrollability. Otherwise, interpretations vary among persons and from one situation to another. Pure outside and non-interpretative observation of human happiness, thus, is not an option. Measurements and socio-technical interventions presuppose the use of specific interpretations as dogmas.

Modern science evolved as a fierce opponent of the dogmatism of scholastic learning. Is there a risk that it might come full circle?

The textual snapshot 'Well-Being Units' draws on Gitte Meyer, Lykkens kontrollanter: Trivselsmålinger og lykkeproduktion [The happiness controllers: The measurement of well-being and the production of happiness]; the quotation is taken from Gus O'Donnell, Angus Deaton, Martine Durand, David Halpern and Richard Layard, 'Wellbeing and Policy'.

classical idea(l) of citizenship.[93] Used as social rather than political concepts, the notions of publics and scientific citizenship – often presented within an intricate framework of natural science metaphors – seem only to have reinforced a view of science as the defining feature of society and political life. Meanwhile, the elitism–populism axis has been left in place as has the consequent understanding of popular communication and the view of society that gave rise to both. It should come as no great surprise, then, that critique of the deficit model keeps bobbing up in the scholarly discourse while in wider contexts its basic condescending assumptions about the general public have remained unaltered and form the basis of widespread science communication routines, connected to understandings of science communication as a mass communication variety.

Fascination as a Science Communication Ideal

In 1528, Thomas More (1478–1535) was asked by the bishop of London to write a text for 'the simple and unlearned'.[94] Surprisingly little has changed. After centuries of educational and enlightenment efforts the writing of texts for the simple and unlearned is a neatly condensed description of current mainstream approaches to science communication, targeting the supposed masses on the basis of widespread and deeply rooted assumptions or social prejudices.

Prominent among those prejudices – irrespective of whether the masses of the people are appreciated as an object of reverence or loathed as an object of contempt – is the assumption that individual units of the masses are disinclined towards abstract thinking while their emotions are easily roused. The features ascribed to them have childlike qualities. The adults, then – the elites – must address them accordingly. To bridge the imagined gap between the masses and the elites, appearing as two separate worlds, representatives of the elites have to apply the art of hitting below the intellect,[95] using criteria such as dramatization, emotional appeal and what's-in-it-for-me approaches.

Against that background it makes sense that fascination is a widespread – possibly *the* most widespread – science communication ideal. 'Automatism, hypnosis, suggestion, hallucination, magnetism, somnambulism, collective hysteria' have been described as 'key words of crowd psychology'.[96] Fascination might be added to the list, as a key word of applied crowd psychology, *and* of widespread understandings of science communication.

Originating in *fascinus*, the Latin term for spell or witchcraft, the literal meaning of 'to fascinate' is to cast or put under a spell.[97] To fascinate an audience, thus, amounts to bewitching that audience, using purely emotional appeal while avoiding appeals to the faculty of reasoning. The strategy of using fascination as a means of – fuel for – the promotion of messages has proved effective often enough and in many contexts. Seen as an aim of science communication it does, however, appear distinctly odd and contradictory. Science, widely taken to constitute the epitome of reason, is equally widely taken to be dependent for its communication on appeals to the quintessence of irrationality; and its preference for facts is assumed to be furthered primarily by appeals to the perceived opposite of facts – feelings. As a consequence, damage to the science cannot be avoided.

Attempts to fascinate mass audiences, presumed to be less than bright, reintroduce or reveal the continued presence of magical elements that the exact sciences were supposed to have been purged of. Paradoxically, this may result in presentations that, because of a stress on methodology as a kind of magic, appear to increase rather than reduce the complexity of issues, in particular when scientific methods are applied to trivial everyday topics.

Science communication, no wonder, is often seen as a rather degrading activity,[98] lacking intellectual rewards – as stated in 1994 by a prominent British scientist referring to his science communication activities: 'I'm a sort of pornographer of science. The role of being a communicator of science is far more ignoble than being a scientist. But someone has to do it.'[99]

On its own contradictory premises, the logic apparently adds up as a model for the transportation of knowledge from knowers to non-knowers, from insiders to outsiders or from pornographers to punters. The latter, we may safely assume, are not licensed to make critical enquiries and ask sceptical questions. That is the prerogative of the former.

During the most recent decades, however, a risk of a boomerang effect has evolved, endangering even sceptical exchanges in communication among scientists. The expansion of science, accompanied by the growth of cross-disciplinary research, has destabilized the twin notions of insiders and outsiders.

Traditionally, a principle of disciplinary autonomy has served as the organizing principle for the exercise of knowledge scepticism in science. The notion of a discipline signifies, by convention, the fundamental unit of science. The disciplines have been regarded as the sources of new and reliable knowledge,[100] and they have served as centres for the maintenance and transmittance of established knowledge. Education and training efforts have grown from and have been directed at the disciplines. Demands for loyalty and adherence to norms have circled primarily around the disciplines. Measures to secure the quality of research results have been anchored within disciplines. Peer review is supposed to be review executed by one's disciplinary colleagues. Metaphorically speaking, the disciplines may be regarded as the nation states of science – and as nation states they do not only regulate internal affairs but are also interrelated in a sort of interdisciplinary diplomacy. There is a principle of disciplinary autonomy.

Across disciplines, scientists are expected to respect the territories of other disciplines, and to respect the interdisciplinary norm that specialized knowledge claims can be reviewed competently only by insiders. There is a norm of non-intervention. In science communication, the autonomy principle has worked for a long time as a barrier to the exercise of (knowledge) scepticism by non-scientists in public exchanges, but the principle of disciplinary autonomy has the potential to prevent even the scrutiny of knowledge claims across disciplines. Cross-disciplinary scepticism and criticism is not only not encouraged but may be actively discouraged as incompetent by definition. While directly affecting the communication between scientists, this even has the potential to indirectly affect science communication in public.

Formally working together as colleagues in multidisciplinary projects, scientists from different disciplines are likely to sometimes regard each other as *fellow scientists* and sometimes as *laymen*. Fellow scientists may be seen exactly as fellow scientists in relation to the public at large. Thus, participants in multidisciplinary research projects may be licensed to make public announcements about the progress and positive expectations of the research, in that context appearing as representatives of science at large. At the same time, the internal communication practice is likely to be guided by a stress on the lay aspects of the perception of scientists from disciplines other than one's own. Seen from this perspective, they appear as non-specialists who should not exercise scepticism in foreign disciplinary territory. Common to the role of the fellow scientist and to that of the layman, in other words, is the feature of *the non-specialist* as related to the traditional principle of disciplinary autonomy.

(Over)simplification also in internal exchanges is a possible consequence – either based on the assumption that the others, in their capacity as fellow scientists, are already aware of the complexities, or on the assumption that the others, in their capacity as laymen, would not be able to appreciate the complexities. Moreover, in relation to the public at large, researchers might borrow authority from each other to speak on behalf of all the disciplines involved in multidisciplinary projects – and with a tendency to (over)simplify not only because a public of laymen is assumed to be unable to appreciate uncertainty and complexity but also because the researchers themselves are unfamiliar with sceptical exercises outside their own disciplines.

This might lead to routines of oversimplification – and perhaps even the use of fascinating appeals – being transferred from science communication directed at the public at large to cross-disciplinary communication. Thereby, in turn, oversimplification in science communication in general may be reinforced, creating even greater difficulties for public exchanges about science-related political issues and, of course, for debates on how to understand such issues in the first place. In what sense are they political and how, then, should they be dealt with? There are different understandings of politics to choose among. That is the topic of Chapter 4, 'The Elusive Concept of Modern Politics'.

Notes

1   *Duden: Das Herkunftswörterbuch*; Ordbog *over det danske sprog* [Dictionary of the Danish language].

2   See for instance Tom B. Bottomore, *Elites and Society*; John Carey, *The Intellectuals and the Masses: Pride and Prejudice among the Literary Intelligensia 1880–1939*; José Ortega y Gasset, *The Revolt of the Masses*; Thorstein Veblen, *The Theory of the Leisure Class*; C. Wright Mills, *The Power Elite*.

3   Chistopher Hill, *The Century of Revolution: 1603–1714*.

4   Albert Hourani notes, referring to a philosopher from the ninth century: 'The distinction between the intellectual élite and the masses was to become a commonplace of Islamic thought.' Hourani, *A History of the Arab Peoples*, 78.

5   Reinhart Koselleck, *Begriffsgeschichten*, 406–08.

6   See for instance, ibid., 402–64.

7   A. S. Hornby, ed., *Oxford Advanced Learner's Dictionary of Current English*. Fifth edition.

8   Quoted in Koselleck, *Begriffsgeschichten*, 424.

9   Ibid., 423–24.

10 My discussion of the concept of the middle classes has been informed by Koselleck, *Begriffsgeschichten*, 412 and 425–26.

11 There are multiple interpretations of the relationship between the concepts of the middle classes and the masses, respectively. Some only include the so-called working classes in the masses; others even include the middle classes.

12 Edmund S. Morgan, *Inventing the People: The Rise of Popular Sovereignty in England and America*.

13 There seems to have been a post-reformation attitudinal move from a preference for wit to a preference for sentimentalism. See James Simpson, *Burning to Read: English Fundamentalism and Its Reformation Opponents*, 256, and Hill, *The Century of Revolution*, 300.

14 Hannah Arendt, 'Kultur und Politik', is one of many possible examples of critical enquiry into the notion of the masses as a qualitative concept.

15 George Bancroft (1800–1891) was an American historian and politician.

16 Gordon S. Wood, *The Radicalism of the American Revolution*, 360.

17 For an account of how quantitative knowledge was connected to understandings of democracy, see Theodore M. Porter, *Trust in Numbers: The Pursuit of Objectivity in Science and Public Life*.

18 The assumption that the intellectual and the popular are opposites is not a universal phenomenon. Scandinavian cultures – Danish and Norwegian in particular – have been much influenced by assumptions of intellectual capacities and leanings among the public in general. Those assumptions, in turn, combined with nationalist emotions, resulted during the late nineteenth and early twentieth century in the rise of hundreds of voluntary schools (*folkehøjskoler*) for youngsters, not least from peasant families, to facilitate their access to learning in the humanist sense.

19 David Collier, Fernando Daniel Hidalgo and Andra Olivia Maciuceanu, 'Essentially Contested Concepts: Debates and Applications'.

20 Gustave le Bon was and probably has remained the most widely known of the early crowd theorists. At the time, however, the gist of his understandings was quite mainstream, and related ideas about the capacities of the masses of the people were expressed in, for instance, Italy and Britain by Scipio Sighele, *Psychologie des Auflaufs und der Massenverbrechen*, and William McDougall, *The Group Mind*, 21–47, respectively.

21 Gustave le Bon, *The Crowd: A Study of the Popular Mind*, foreword.

22 Ibid., introduction. Le Bon also connected barbarism to instincts: bk. I chap. I, bk. II sec. 3.

23 Ibid., bk. I chap. I, bk. I chap. II, bk. I chap. II sec. 4.

24 Ibid., bk. I chap. II.

25 Ibid., bk. I chap. II sec. 2.

26 Ibid., bk. I chap. II sec. 3.

27 Ibid., introduction, bk. II chap. IV sec. 2.

28 Ibid., bk. I chap. III sec. 1.

29 Ibid., bk. I chap. III sec. 2.

30 Veblen, *The Theory of the Leisure Class*, 144.

31 Robert K. Barnhart, ed., *Dictionary of Etymology*.

32 Veblen, *The Theory of the Leisure Class*, 18, 20, 21, 15, 3, 15 and 41.

33 Ibid., 15, 16, 17, 92.

34 Ibid., 109, 130–31, 98, 118, 112–13, 121.

35 Ibid., 97, 80–84, 63, 65, 128.

36 Ibid., 144–58, 144, 145, 147, 146.

37 Ibid., 148, 151, 153.

38 Literally, *otium cum dignate* means free time with dignity, or the use of free time in a dignified way. The term *otium* (from the Latin), like the term *school* (from the Greek), originates in a term for free time as opposed to time that has to be spent, slave-like, on the production of necessities – labour. The term was coined in a context where such free time was highly valued and connected to human dignity. In contrast, the current notion of the *otiose*, in line with Veblen's understandings, is used to characterize activities as a (non-productive) waste of time. Barnhart, ed., *Dictionary of Etymology*; Hornby, ed., *Oxford Advanced Learner's Dictionary of Current English* (eighth edition); *Ordbog over det danske sprog* [Dictionary of the Danish language].

39 Veblen, *The Theory of the Leisure Class*, 154, 155–56.

40 Hill, *The Century of Revolution*, 67, 85, 207, 239.

41 Ibid., 101.

42 Morgan, *Inventing the People*, 92.

43 H. G. Wells, G. P. Wells and Julian Huxley, *The Science of Life: A Summary of Contemporary Knowledge about Life and Its Possibilities*, vol. 1, 2; vol. 31, 973.

44 Barnhart, *Dictionary of Etymology*.

45 Roger Cooter and Stephen Pumfrey, 'Separate Spheres and Public Places: Reflections on the History of Science Popularization and Science in Popular Culture'; Maureen McNeil, 'Between a Rock and a Hard Place: The Deficit Model, the Diffusion Model and Publics in STS'.

46 Barnhart, *Dictionary of Etymology*.

47 Ibid.

48 See for instance, Carey, *The Intellectuals and the Masses*, 141, 144, 147.

49   Wells et al., *The Science of Life*, vol. 31, 969.

50   Ibid., vol. 31, 975–76.

51   Quoted in Leon Fink, *Progressive Intellectuals and the Dilemmas of Democratic Commitment*, 41.

52   Hermann J. Muller, *Out of the Night: A Biologist's View of the Future.*

53   Julian Huxley, *Memories*; Julian Huxley, *Memories II.*

54   Hermann J. Muller, 'Genetic Progress by Voluntarily Conducted Germinal Choice', 254–56.

55   Gordon Wolstenholme, ed., *Man and His Future*, 294–95, 282, 290.

56   Joshua Lederberg, 'Biological Future of Man', 271.

57   The early populists took the people to belong outside the sphere of particular interests. See for instance Jeff Ludwig, 'From Apprentice to Master: Christopher Lasch, Richard Hofstadter, and the Making of History as Social Criticism', 4, 18, 19, 27.

58   An early version of this argument appeared in Gitte Meyer, 'Scientists, Other Citizens, and the Art of Practical Reasoning'. It was fully developed in Gitte Meyer, 'In Science Communication, Why Does the Idea of a Public Deficit Always Return?'.

59   Thomas Sprat, *History of the Royal Society*, 113.

60   Barnhart, *Dictionary of Etymology.*

61   John Dewey, *The Public and Its Problems.*

62   Walter Lippmann, *Public Opinion.*

63   Dewey, *The Public and Its Problems*, 215.

64   Ibid., 174, 221–32.

65   Lippmann, *Public Opinion*, 48.

66   Ibid., 2–20, 195.

67   Ibid., 250, 254.

68   Fink, *Progressive Intellectuals and the Dilemmas of Democratic Commitment*, 18–19.

69   Quoted in J. Herbert Altschull, *From Milton to McLuhan: The Ideas Behind American Journalism*, 308.

70   Contemporary critics of the assumptions that form the bases of the elitism–populism axis did exist. In 1921, for instance, literary critic Harold Stearns was not impressed by 'propaganda-experts' who used as their point of departure 'what they thought to have discovered as the infinite docility and suggestibility of the mob'. Stearns, *America and the Young Intellectual*, 93.

71   José Ortega y Gasset, *The Revolt of the Masses*, 67, 82, 85, 90, 100, 109, 110–11.

72   Ortega y Gasset may have had the French rather than the British Enlightenment in mind.

73   Ortega y Gasset, *The Revolt of the Masses*, 111.

74   Ibid., 112.

75   Philip E. Converse, 'The Nature of Belief Systems in Mass Publics (1964)', 43, 6.

76   Ibid., 10.

77   Ibid., 10–11.

78   Ibid., 10–11, 15.

79   Ibid., 13, 15.

80   Ibid., 17, 27–28.

81   Ibid., 34, 69n22, 35, 38–39, 67n13, 46, 54.

82   Ibid., 68–69n21.

83   Seymour Martin Lipset, *Political Man: The Social Bases of Politics*, 103, 107.

84   Ibid., 109–19, 120, 130n75, 122–23.

85 It should be kept in mind that the writings of Converse and Lipset were probably influenced by the contemporary fear of a communist threat. The assumptions and conclusions they represented have, however, had long-term consequences in a much wider context.

86 Fink, *Progressive Intellectuals and the Dilemmas of Democratic Commitment*, 3–4.

87 Gitte Meyer, *Offentlig fornuft? Videnskab, journalistik og samfundsmæssig praksis* [Public reason? Knowledge, journalism and societal practice].

88 See for instance LeeAnn Kahlor and Patricia A. Stout, *Communicating Science: New Agendas in Communication*; and Frank Zenker, 'The Explanatory Value of Cognitive Asymmetries in Policy Controversies'. In 2007, the deficit model controversy was described as a dividing line between opposing camps in the science communication discourse; see Edna Einsiedel, 'Editorial: Of Publics and Science'.

89 Patrick Sturgis and Nick Allum, 'Science in Society: Re-evaluating the Deficit Model of Public Attitudes'.

90 Martin W. Bauer, 'The Evolution of Public Understanding of Science: Discourse and Comparative Evidence'. Bauer's italics.

91 Ian Welsh and Brian Wynne, 'Science, Scientism and Imaginaries of Publics in the UK: Passive Objects, Incipient Threats'.

92 Maja Horst, 'Public Expectations of Gene Therapy: Scientific Futures and Their Performative Effects on Scientific Citizenship'.

93 Ibid. On the notion of scientific citizenship see also, for instance, Ulrike Felt, ed., O.P.U.S. Optimising Public Understanding of Science and Technology: Final Report.

94 Quoted in Simpson, *Burning to Read*, 238.

95 The expression of hitting below the intellect has been widely attributed to Oscar Wilde (1854–1900) who, of course, invented it in another context.

96 Armand Mattelart, *The Invention of Communication*: 249.

97 Barnhart, *Dictionary of Etymology*.

98 Stephen Hilgartner, 'The Dominant View of Popularization', noted the understanding of science communication as a degrading activity in 1990.

99 Quoted in Gitte Meyer, 'Expectations and Beliefs in Science Communication: Learning from Three European Gene Therapy Discussions of the Early 1990s'.

100 John Ziman, *Reliable Knowledge*.

Chapter 4

THE ELUSIVE CONCEPT
OF MODERN POLITICS

Just about a century ago, the concept of technocracy was introduced into the English language as a positive term signifying an orderly society, governed like a well-oiled machine by an elite of technicians. In current European usage, the concept is mostly used as a term of abuse, but the inventor of the term – British-born American engineer William Henry Smyth (1885–1940) – was blessed with ignorance of that future fact. He introduced the concept into American usage in the wake of the Great War (1914–18). Calling for 'a Supreme National Council of Scientists – supreme over all other National Institutions – to advise and instruct us how best to Live, and how most efficiently to realize our Individual and our National Purpose and Ideals', he described the members of this supreme council as the 'Managing Directors' of society.[1]

In the United States, according to Smyth, the war had facilitated the development of a completely new form of government. He called this new form technocracy, defining the meaning of the term as 'the organizing, co-ordinating and directing through industrial management on a nation-wide scale of the scientific knowledge and practical skill of all the people who could contribute to the accomplishment of a great national purpose'.[2]

Smyth was a visionary. His vision and mission concerned the replacement of politics by scientific management. At the same time, however, he saw himself as a proponent of democracy: 'Carry this new form of government into the days of peace,' he argued, 'and we will have industrial democracy – a new commonwealth.' Indeed, to his mind, technocracy was similar to 'rationalized Industrial Democracy'. It would also, as he saw it, be a significant improvement if compared to democracy 'in the rough' – that was, in the shape of 'the rule of the mob, the rule of the masses, the rule of the majority – the rule of un-intelligence'. Humankind, according to Smyth, was driven by 'the four great human instincts – to live, to make, to take, to control'[3] and therefore needed a superior technical-scientific intelligence to rule it.

Smyth's vision is a simplistic example of widespread understandings of politics. Using the logic of science as their yardstick, they all belong in a dualistic scheme of thought and rely for the making of distinctions on such supposed dichotomies as truth versus power, facts versus feelings, knowledge versus values and rationality versus irrationality. Politics is taken to be either the irrational opposite of science as the epitome of rationality, or it is taken to be the rational application of science. None of these understandings, different as they are in many ways, represents a substantial idea of politics as an independent activity in its own right.

Understandings of politics as the opposite or the application of science are mutually exclusive. As a consequence, it would seem possible to get rid of politics in the former sense by introducing the latter variety. That was the purpose of Smyth's advocacy of technocracy and even though today the specific term is mostly used as a term of abuse, the basic assumptions that furthered its introduction as a positive term have remained forceful. A range of current social-instrumental practices – approaches to science communication included – would likely have been praised by Smyth as essentially technocratic. Even more common, probably, is the use of politics as a term of abuse, denoting grubby mixtures of power play, corruption, highly strung emotions, and partisanship. All vice, apparently, belongs in the realm of politics unless it surrenders to science.

It is difficult to get farther away from the classical, Aristotelian understanding of politics as a distinct and distinctly human endeavour in its own right, allowing humans to unfold their capacities for thought and speech and, thereby, to cope with the uncertainties of the human condition and create a civil society. Perceiving exchange among different points of view to be pivotal to political life, this was a pluralistic understanding of politics. It was accompanied by an idea of political reasoning as the highest and most worthy form of practical reasoning – *phronesis* – as distinct from the twin founding concepts of modern science: *techne*, technical reasoning, and *episteme*, the contemplation of universal truths.

Classical thought has been hugely important to the development of modern Western thought, ideals and strivings. At the same time, the attitudes towards politics have moved in a completely different direction. As a rule, politics is no longer ascribed qualities of its own. Defined as the rational application of science, it may be tied to the state and its technical systems of regulation and administration. Defined as the irrational opposite of science, it is linked either to cynical power plays and deceit or to ideological warfare between extremes. How may that change of direction have come about? And, how do understandings of politics that use science as their yardstick affect science communication routines and models of thought vis-á-vis the growing number of science-related political issues?

Suspicion

A certain amount of mutual trust among citizens as peers was presupposed in Aristotle's political thought. His stress on the capacity for speech and on exchange among different points of view would not, for instance, make much sense if participants could not be relatively certain that other participants were, as a rule, inclined to speak truthfully and make reasonable judgements. In contrast, suspicion has been a frequent feature of modern political thought and cultures, so much so that there is a tendency to portray modern political cultures marked by a low degree of suspicion as havens of complete trust and, thus, curious aberrations from a supposedly natural order of suspicion.[4]

Religious fanaticism, persecution and civil wars have been drivers of mutual suspicion, leaving those involved with experiences that have influenced the evolving political cultures. Civil wars, in particular, have left their marks in terms of fear of conflict and disagreement – as lack of trust, that is, in the capacity of fellow humans to deal with conflicts in civilized ways. And most probably, the rise of the marketplace, with its stress on competition, has had related effects, making society appear as a battleground where everyone fights against everyone else to assert their own particular interests. Whatever the causes – they are undoubtedly exceedingly complicated – the trait of suspicion is a fact of modern political cultures, although more dominant in some than in others, and has influenced even the logic of modern science with its wariness of human interpretations and judgements as potential sources of partiality and error.

The trait of suspicion appears to be – and to have been – particularly deep-rooted and taken for granted in the United States. American historian Theodore Porter refers to 'the American political context of systematic distrust',[5] and he discusses the trust in calculations as an effect of 'the corruption of politics' and as 'one scheme for neutralizing politics'.[6] Looking back to the end of the eighteenth century, Gordon Wood, another American historian, notes: 'People increasingly felt so disconnected from one another and so self-conscious of their distinct interests that they could not trust anyone different or far removed from themselves to speak for them in government. American localist democracy grew out of this pervasive mistrust.'[7] And yet another American historian, Richard Hofstadter (1916–1970), characterized the style of American politics as paranoid. American 'political psychology', he found, was pervaded by 'heated exaggeration, suspiciousness, and conspiratorial fantasy', by 'eschatological' ideas and an 'apocalyptic and absolutistic framework'. The 'paranoid', he commented, 'is a militant leader. He does not see social conflict as something to be mediated and compromised, in the manner of the working politician. Since what is at stake is always a conflict

Open-Mindedness or Raving Madness?

When, in 2005, I asked European bioscientists about the possible use of reproductive cloning in humans, two interviewees, representing the same scientific speciality, disagreed strongly.

'To produce children simply for one's own sake', one interviewee responded, 'that is the most offensive thing I can think of. Will they develop cancer at 30? Will they be affected by high blood pressure at 20? Our calves [produced by way of various reproductive techniques] develop diabetes. It is extremely unnatural in ruminants. It is almost impossible, under normal conditions, to provoke diabetes in ruminants. [...] To use cloning techniques on humans, one would have to be raving mad.'

Another interviewee had it that 'as a scientist you cannot say no to a possibility. As a priest or a nun you can say no, but not as a scientist. I must try to keep an open mind. [...] Human reproductive cloning might become possible in ten years, perhaps.'

Scientists may disagree like other people, even on knowledge-related questions. Why is that widely considered a cause for embarrassment?

The textual snapshot 'Open-Mindedness or Raving Madness?' draws on Gitte Meyer, *Why Clone Farm Animals? Goals, Motives, Assumptions, Values and Concerns among European Scientists Working with Cloning of Farm Animals.*

Model Politicians

For a long time, a Danish economist told me in 2003, economists 'modelled politicians as people who were concerned with the common good. Today, we have moved to a completely different playing field, assuming that politicians are only out to feather their own nests. Assessing models, you do of course need information about what kinds of assumptions they have been based upon'.

The textual snapshot 'Model Politicians' is a translation from Gitte Meyer, *Offentlig fornuft? Videnskab, journalistik og samfundsmæssig praksis* [Public reason? Knowledge, journalism and societal practice], 290.

between absolute good and absolute evil, what is necessary is not compromise but the will to fight things out to a finish.'[8]

Politics, in particular, is and has been viewed with suspicion. Throughout the Progressive Era, according to British political theorist Bernard Crick (1929–2008), the term 'political' was 'largely held in contempt'.[9] Along related lines, American historian Edmund S. Morgan (1916–2013) used the term 'politician' with reference to 'the pejorative connotations that the word has always carried'.[10] Those connotations probably explain why political as a term is frequently used to signify the dark sides of human, social life but is substituted by the term 'social' when references are made to the brighter aspects of social relations.[11]

The suspicion of politics is, however, older than the United States. Although it may be particularly emphasized in the United States, it is not a purely American trait.

The highly influential British philosopher Thomas Hobbes (1588–1679) was one of the earliest thinkers to, in the seventeenth century, express the suspicion on a grand scale. The assumption that warfare was the natural state of affairs among humans formed the basis of his political philosophy. Perceiving humans as social beasts, he found them even worse than the animal kingdom at large. That was so because humans were prone to having and to vigorously defending ideas and opinions and to exercise presumptuous private judgement. Therefore, agreement to absolute obedience to a sovereign was necessary for the protection of peace. Like many other contemporary thinkers, Hobbes was deeply affected by the brutalization of English society by civil wars and he considered his sketch of a strongly authoritarian society, governed by fear, a necessary evil, not a Utopia.[12]

The lasting intellectual consequences of the English civil wars, polarizing the nation, it has recently been argued, were 'those which turned against, or stood back from, the passions that had animated the conflict'.[13] The sometimes passionate fear of those passions, in turn, was connected to distrust in humans as social animals and to the assumption that humans were driven by mutual fear.

In a discussion of the violence following in the wake of the Reformation, Australian American historian James Simpson has noted: 'The collapse of faith in the secular realm produces extreme commitments to religious faith. It is precisely when faith in the conduct of political life has collapsed that revolutionary spiritualities of "faith alone" rise rapidly and aggressively to prominence.'[14] Passionate beliefs in science as a substitute for politics might be generated in much the same way. Suspicion of human interpretations and judgements – unavoidable activities to participants in social and political life – may inspire the belief that the position of the outside observer is not only the safest position but also the only position that facilitates the generation of reliable knowledge, taken to be free from human interference in the shape of interpretations and judgements.

The negative perception of the social life of humans as belonging to a sphere of power relations, self-interest and inequality has a positive and optimistic counterpart, celebrating a social sphere of original unity and harmony, composed of humans who – because they all share the same animal qualities[15] – are fundamentally equal. Maintaining the social perspective with its view of humans as social beasts on a par with other such beasts, inequality and division into social groups may be seen as unnatural distance, as a distortion of a fundamental unity of humankind. It may even be considered a fall from grace. From that point of view, then, the (hi)story of human social relations and communication begins with closeness, intimacy and trust rather than warfare.

In 1776, Thomas Paine (1737–1809), for instance, told this other story in *Common Sense*. He imagined 'a small number of persons settled in some sequestered part of the earth' and took them to represent 'the first peopling of any country, or of the world'. To this small group of persons, seeking assistance and relief from one another, social relations – society, in Paine's term – represented a state of 'natural liberty' and 'reciprocal blessings', and when they first assembled under a tree 'to deliberate on public matters', there was no representation of anybody by somebody else, but: 'In this first parliament every man, by natural right will have a seat.'[16]

The social perspective, in short, which has informed modern understandings of politics and many modern political practices, was born with ambivalence and evolved within a framework of assumed dichotomies. Schematically put, pessimistic versions are based on an assumption of original warfare between humans, whereas optimistic versions are based on an assumption of original unity and harmony. Accordingly, one version sees the exchange of ideas and opinions as an instance or precursor of warfare; whereas the other version takes communication – in the shape of dialogue – to be a means to maintain or restore unity. Varieties of those opposite understandings have coexisted and been at odds for centuries. They represent one of the fundamental tensions of modern societies.

Either way, whether fearing or celebrating humans as social animals, they all seem to have been marked by a rather strong aversion to any form of politics that goes beyond the local face-to-face gathering. From one position, politics appears as the expression of a social sphere of power play and partisanship among self-interested parties. From another position, politics – as power play and partisanship among self-interested parties – destroys a social sphere of unity, harmony and intimacy. Paine found that '[s]ociety is produced by our

Vaccination and Polarization

For about two centuries – the first smallpox vaccination was ready in 1798 – vaccination has been a divisive issue, subject to polarized discussions between general pro- and anti-vaccination positions. Anti-vaccination societies were formed in the early twentieth century and, characteristically, vaccination debates have rarely been calm enquiries into the possible benefits and costs, to individuals and to society as a whole, of specific vaccines. Rather, having created a pattern for exchanges on other science-related public affairs, vaccination debates have often played out as ideological battles between seeming representatives of pro- and anti-science attitudes.

But how is it possible to be a proponent or opponent of vaccinations in general? Each case comes with a wide spectrum of issues that need attention: the nature and seriousness of the disease, the effectiveness of the vaccine, uncertainties about possible side effects of the vaccination, how it is produced, financial and other motives behind the introduction, et cetera. Even when such issues have been looked into there may still be reasonable pros and cons to consider, and it is perfectly possible to reason one's way to being in favour of some but dismissive of other vaccination schemes. Why, then, are people almost forced into general pro- versus anti-vaccination stances, taken to mirror general pro- versus anti-science attitudes, each time a new vaccine is developed?

Historically, vaccination battles appear to have been particularly fierce when connected to compulsory vaccination schemes; common in some but rare in other cultural contexts. The threat of the forceful application of science – involving science as an authority in a very direct sense in a seemingly fundamental conflict between each individual and society – may have contributed to preparing the ground for today's polarized debates on vaccination and other science-related political issues.

For background information of relevance to the textual snapshot about vaccination debates, see, for instance, College of Physicians of Philadelphia, *The History of Vaccines*; and Jason L. Schwartz, 'New Media, Old Messages: Themes in the History of Vaccine Hesitancy and Refusal'.

wants, and government by our wickedness; the former promotes our happiness positively by uniting our affections, the latter negatively by restraining our vices. The one encourages intercourse, the other creates distinctions.'[17] Hobbes used a game of cards as his favourite metaphor for politics.[18]

The logic of modern science, in contrast, with its idea(l) of impersonal outside observation, can be partly seen as an alternative to suspicious politics.

The Opposite or the Application of Science

Images of politics as the wicked opposite of science come in different shades, with different emphases.

Using the assumed dualism of observation versus participation as its point of departure, one version links politics to partisan participation. Politics, then, becomes reminiscent of religious civil wars, complete with crusaders fighting against other believers from extreme and irreconcilable positions of a radical or fundamental nature. Polarization is seen as the natural (dis)order of politics, viewed as a highly emotional, moralistic and irrational kind of activity. A whole array of other assumed dichotomies contribute to shaping this understanding of politics – thought versus action, the intellectual versus the emotional and brain versus body among them.

As outside observation, thought, intellectual activity and brain are attributed to science; politics is left with participation as partisanship, action, emotions and body. This is old. Clearly it was a vintage idea of politics when, in the late nineteenth century, French sociologist Gustave le Bon (1841–1931) referred in passing to 'every thing that belongs to the realm of sentiment – religion, politics, morality, the affections and antipathies, &ce.'.[19] He was in line with present understandings of political participation as the fervour, shouting, cheering et cetera of a group of people gathering around a common cause.[20] Importantly, this image of politics ties political activity to ideologies and to battles between directly opposed ideologies, only distinguishable from religions by the absence of actual deities. Normatively, politics in that sense may be valued negatively, as dangerous and potentially explosive, or positively, as a necessary supplement – enthusiastic and warm-hearted – to cool-headed science.

Yet another, no less widespread image of politics has it that politics constitutes the epitome of cynicism and power plays and that politicians are merely self-serving elitists. Typically, in English, 'cynicism' is one of those disapproving terms that seem to be most easily explained by references to politics.[21] Operating on the basis of the assumed dichotomy of truth versus power, this understanding takes the true and the good to be of a kind and to be residing in science. Politics is tied to power but not necessarily to the irrational.

Rather, it is connected to rational calculation that does not serve the common good but only particularistic self-interest. Morality, thus, is placed on the side of science which is supposed to be using calculations to serve the common good and to be eminently suited for doing just that because of its chosen position as an outside, non-partisan observer.

In both cases, the notion of partisanship is pivotal. Whether politics is viewed as completely emotional or as completely devoid of emotions that go beyond self-interest, it is taken to be marked – and separated from science – by partisanship.[22] It is a stark term, signifying an understanding of political participation as aggressive, intemperate and one-sided. Whether seen as wicked or virtuous, it is far removed from idea(l)s of reasonable and moderate participation.

Although they have a history – local and specific – of their own, these reduced images of politics have gained momentum on a global scale. So has the positive understanding of politics as the application of science. With its emphasis on the impersonal, science can be seen as the ideal alternative to the kind of exercise and abuse of personal power by hereditary, landowning aristocracies that were shaping the almost feudal order of the day when modern science was founded. Scientific management and administration is, in principle, committed to the impersonal exercise of power. Its prime concern is the implementation of policies, and policies – as distinct from politics – represent decisions that have been made and are not up for discussion. As of today, more often than not, such policies are or are pretended to be science-based.

The understanding of politics as the application of science is connected to a purpose – consciously or unconsciously, tacitly or overtly – of getting rid of politics altogether. The most extreme versions aim at full technocracies suited to managing societies that are seen primarily as economies. Politics – or rather, policies – are seen as means to support the sphere of production and may even be viewed as products themselves. It becomes the task of politics to take control and solve technical problems rather than to deal with practical challenges.[23] Indeed, all societal problems appear to be of a technical nature.

It has come naturally, in continuation of those understandings of politics that use science as their yardstick, to also make politics an object of scientific enquiry. Bernard Crick characterized this as a particularly American approach to politics and, in 1959, dedicated a treatise to *The American Science of Politics*, seeking to 'explain the special plausibility to American students of politics of the view that politics can be understood (and perhaps practised) by "the method of the natural sciences"'.[24] Along these lines, American political scientist Robert Dahl (1915–2014) wrote, seemingly without hesitation, about 'political scientists and other technicians'.[25]

Crick was not positively impressed by the approach and found that 'enthusiasm and persuasive zeal' and 'intense democratic moralism' were hallmarks of the outcomes of the scientific efforts, making 'the whole school [appear] more as an expression of American political thought than of science'.[26]

As frameworks of thought originating in the United States have been adopted all over the world, the idea of a science of politics has lost any flavour of Americana it might have had. It has become a global challenge to ponder the possible effects of applying purportedly non-normative scientific methods to the study of complex issues with normative elements, attempting to circumvent normativity by way of denial. That denial, in turn, rather than dissolving the normative elements of the issues may serve only to chase them out of reach for critical thought, allowing them to morph into unrestrained moralism.

It would seem there is a possible lesson here for students of science communication: when dealing with human affairs, such as science communication, normativity cannot be circumvented but has to be confronted directly. Rather than constituting a tainted substitute for pure and non-normative description of mechanisms, the open exercise of (personal) judgement and (civilized) exchange among different points of view can be seen as benign alternatives to allowing normativity to run amok and turn moralistic.

The dominant understandings of politics, because they use science as their yardstick, do not facilitate the learning of lessons along such lines. They are tuned, rather, to inspiring attempts to develop science communication into a small-time science in its own right – a science, that is, intended not least to further democracy. And which, thereby, unwittingly exemplify the surprising phenomenon of anti-political devotion to democracy.

Anti-political devotion to democracy

To those who take democracy to be an out and out political concept, the widespread occurrence of anti-political devotion to democracy, combining devotion to democracy with aversion to politics, *is* a surprising phenomenon. It makes sense, however, if democracy is seen in the first place as a social concept connected to views of politics as a social phenomenon. From this viewpoint, status relations become pivotal to the understanding of democracy.

Classical political thought took citizens to be political equals by definition and was, for a number of reasons, relatively indifferent to social (in)equality. The political equality of citizens was a foundational cornerstone; it was a premise of politics, as was the understanding that politics was a participatory activity, not merely presupposing but actually defined by civic activity.[27] In contrast, the social *in*equality of citizens is a premise of democracy as a social

Snapshot XIII

The Mental Climate of the Climate Debate

'Alarmists', 'hysterics', 'doomsayers', 'dissenters', 'sceptics', 'deniers'. The mental climate of the debate on human-made climate change was anything but civil in the first decade of the twenty-first century. It was a polarized debate between two opposing camps, both equally convinced – the 'deniers' no less than the 'alarmists' – that they had science on their side while the opposing camp was caught up by religion, by vested interests or, worse, by politics. Their claims were fabricated; they were forging the truth; they were 'politicizing' science.

The debate has moved on. In wide areas of Europe, at least, the majority public opinion probably matches the majority opinion among scientists in the field of climate research that current climate changes are to a large extent human-made. There is hardly a threatening army of deniers to fight down. Nevertheless, the debating climate has remained fierce. Spurred, it appears, by the 2016 presidential election in the United States, the debate has become a container for all kinds of controversy – immigration policies among them – whether substantially related to the issue or not.

The debate has been turned into a series of crusades. The issue has become a symbol. As such, and sidetracking other environmental concerns, it is used in a battle between proponents of different technology regimes – outmoded and new model production and consumption, respectively – that are equally committed to goals of everlasting and accelerating economic growth.

With its twin ties to science and politics, modern environmental research not only made science play a key role with regard to enquiries into the dark sides of science-based industrial processes, but has also inspired critical discussions about continuous economic growth as a societal ambition. The latter aspect, however, although substantially significant to the issue, has become conspicuously absent in the climate debate.

The textual snapshot about the mental climate of the climate debate has drawn on Gitte Meyer and Anker Brink Lund, 'Klimadiskussionens diskussionsklima: Polarisering i den offentlige debat om klimaændringer' [The debating climate of the climate debate: Polarization in the public debate about climate change].

concept, and the achievement of social equality is its principal aim, closely tied to the goal of furthering the participation of citizens in politics. Inclusion is an aim because most citizens are taken to be excluded from participation in the first place. Being turned into goals, equality and participation are annulled as premises.

The goal of achieving social equality encompasses, among other things, equal and unhampered access to material goods, to education, to information and to making oneself heard in the public arena.[28] Freedom of expression is important to the latter goal because it safeguards the equal rights of all autonomous citizens to speak out in public, rather than because it serves to ensure that a multiplicity of points of view may contribute to enquiries into public affairs. Knowledge of reality is left to science anyway. Thus, the significance of the freedom of expression is not so much that citizens may have substantial contributions to make to shared enquiries, but that each has equal access to say whatever he or she likes without the interference of middlemen and dubious interpreters, and to seek social recognition from others.

Public utterances, on the other hand, should also be accessible to all. From the viewpoint that society is divided into the masses and the elites and that the masses are characterized by a lack of intellectual capacities and leanings, the presentation of intellectual challenges in public communication comes to be perceived as an elitist and undemocratic threat to equality. Wisdom of a popular and democratic nature is tied instead to impersonal numbers.[29] And public opinion – the apparent opinion of the greatest number, the majority – becomes an object of worship.

American statistician George Gallup (1901–1984) had political ambitions along those lines of reasoning when, in 1935, he founded the American Institute of Public Opinion. Thanks to scientific opinion polls it would soon be possible, he argued in 1938, to establish at any time and with the utmost reliability the will of the majority of the people; thereby allowing democracy to reach its highest stage of development. The outcomes of opinion polls, he argued, could be seen as a mandate from the people to their political leaders. Indeed, they could even be seen as means to establishing a modern variety of the classical Greek citizens' meetings, now encompassing the whole of the United States. Public opinion, documented by opinion polls, according to Gallup, constituted a possible counterweight to the separation of political leaders from the people. To Gallup at that time – 10 years later he decided to focus the activities of his institute on market polls – opinion polls were possible means to the end of compensating for increasingly weakening participation in representative democracy.[30]

Such crises of democracy constitute a recurring phenomenon of modern democracies. At different points in time, the disturbing phenomena of political apathy and decreasing participation, accompanied by complaints about the corruption of politics, have been recognized and interpreted as symptoms of a completely new and hitherto unseen crisis of democracy.[31] It might be more helpful to view it as a chronic crisis, erupting now and again and originating in the image of politics in general and representative systems in particular as corrupt by definition.[32]

The pleonasm of 'popular democracy', denoting direct democracy – no middlemen between the people and power[33] – but also connected to the idea of public opinion, belongs in this scheme of thought on democracy along with its counterpart, 'elite democracy'.[34] The latter term signifies and indicates an unfavourable view of representative democracy as an inadequate means of safeguarding the right of the masses of the people to control the political elites, masters or power holders. The introduction of mechanisms to ensure precisely that, in turn, is seen as the raison d'être of democratic systems.

Evidently, these understandings of democracy – with their stress on the impersonal, on numbers and the introduction of mechanisms, and with their aversion to middlemen – bind well together with the logic of modern science. They create a framework that intimately ties science and democracy to each other. Impersonal and quantitative science comes to be seen as democratic almost by definition, compounded by the fact that science also appears as the progenitor of material progress and solutions that make life easier for all and give more people access to more goods. The application of science, indeed, seems to represent a benign and democratic variety of politics.[35]

British scientist turned political theorist John Desmond Bernal (1901–1971) put it the following way in *The Social Function of Science*, published in 1939: 'The scientist may, and indeed must, become a politician, but he will never become a party politician. He sees the social, economic and political situation as a problem to which a solution must first be found and then applied not as a battle-ground of personalities, careers, and vested interests.'[36]

Such understandings of democracy and benign politics do, of course, have bearings on idea(l)s of science communication. Insofar as knowledge is taken to be a form of power – that is generally the case[37] – science communication has a democratic obligation to disseminate knowledge to all as a means of power sharing.[38] Science communication should be aimed at including the masses, addressing them in ways they are assumed to be susceptible to while keeping clear of the suspicious sphere of opinions. Assigned the task of providing knowledge of reality – serving, if you like, as the universal light of humankind – science functions indirectly as a key political player. Paradoxically, scientists are seen as trustworthy politicians because they are assumed to be apolitical.

Sociocracy: More democratic than democracy?

Now and again some have wished to take the political role of science a step further and to put scientists directly in charge of societies, substituting in a straightforward way scientific management and social engineering for politics. The technocracy movement was one example. It was matched by the sociocracy movement.

The idea(l) of sociocracy is roughly similar to that of technocracy but comes with a particular emphasis on the assumed benign influence of technocrats from the social sciences. The term was created by Lester Frank Ward (1841–1913), American paleobotanist turned sociologist, to signify government by social scientists.

According to Ward,[39] there was 'one form of government that is stronger than autocracy or aristocracy or democracy, or even plutocracy, and that is sociocracy'. Neither pleased with majority rule nor with pluralism, Ward saw society as a whole and wanted to enable it to act 'for itself'. To enable it to do so, the body of society, as he saw it, needed a brain in the shape of social scientists.

He suggested that society should

> imagine itself an individual, with all the interests of an individual, and becoming fully conscious of these interests it should pursue them with the same indomitable will with which the individual pursues his interests. Not only this, it must be guided, as he is guided, by the social intellect, armed with all the knowledge that all individuals combined, with so great labor, zeal, and talent have placed in its possession, constituting the social intelligence.

Lamenting the partisanship and corruption of politics, Ward found that 'in the factitious excitement of partisan struggles where professional politicians and demagogues on the one hand, and the agents of plutocracy on the other, are shouting discordantly in the ears of the people, the real interests of society are, temporarily at least, lost sight of, clouded and obscured'.

Social scientists, however, might find solutions to those kinds of problems. Their intervention would ensure that 'the important objects upon which all but an interested few are agreed will receive their proper degree of attention, and measures will be considered in a non-partisan spirit with the sole purpose of securing these objects'.

The investigation of any issue should, of course, be 'disinterested and strictly scientific'. Society or, rather, its social intelligence, 'would inquire in a business way without fear, favor, or bias, into everything that concerned its welfare, and if it found obstacles it would remove them, and if it found opportunities it would improve them'. It was the special characteristic of sociocracy, resting 'directly upon the science of sociology, to investigate the facts bearing on every subject, not for the purpose of depriving any class of citizens of the opportunity to benefit themselves, but purely and solely for the purpose of ascertaining what is for the best interests of society at large'.

Ward was all for the shedding of light on problems from many sides, but, he emphasized, 'in order really to elucidate social problems it must be the dry light of science, as little influenced by feeling as though it were the inhabitants of Jupiter's moons, instead of those of this planet, that were under the field of the intellectual telescope'.

'The great demand of the world is knowledge', declared Ward as a prologue to airing a utopian science communication vision: 'The great problem is the equalization of intelligence, to put all knowledge in possession of every human being.'[40]

Ward and his contemporary Progressives, it has been noted, adopted the identity of intellectual leaders of the crowd and apparently did not appreciate how 'the colonial metaphor of teaching the "natives" how to behave' influenced their efforts and 'however rationalized within prescriptions of neutral expertise, insinuated itself into a discourse of neighborliness, self-sacrificing service, and social partnership'.[41]

It seems to have been the unspoken overall goal to overcome the assumed power versus truth dualism, not by acknowledging truth and power as substantially different but by giving power to truth and its technical equivalent – correct solutions – thereby presumably liberating society from all the dark, political aspects of power.

Visions of revolutionary science

About half a century after the heyday of the Progressive Era, the student movements of the 1960s and 70s called enthusiastically for more politics, democracy and participation. More often than not, however, their understandings of politics and democracy remained tied to science. *The Port Huron Statement* of 1962, issued by the American Students for a Democratic Society, 'yielded a vision informed by a democratic American radicalism going back to Tom Paine'. This was the conclusion of Tom Hayden and Dick Flacks, looking back in 2002 at the statement they had contributed to writing 40 years previously.[42] In continuation of the crusading tradition of politics, the statement used the term 'radicalism' – that usage, actually, is still frequent – with positive connotations to signify the enthusiastic and staunch commitment to a good cause, and the willingness to identify and fight down its enemies.

At the outset, the calls for more politics seem to have been primarily calls for politics as the opposite of science – strongly enthusiastic, emotional and partisan struggle among ideologies with a more than superficial likeness to religious confessions. The students saw apathy, cynicism, hierarchies, manipulation, silence and sterility around them and called for their opposites – activism, authenticity, community, creativity, equality, hope, love, meaning, participation, passion, spontaneity, visions and utopias.

Inspired, not least, by John Dewey (1859–1952) – one of those Progressives who was a direct source of inspiration for the movements – some preferred a view of politics as a possible means to achieve social intimacy, unity and harmony, allowing society to evolve into a great community, relying on the application of science.

The universities were taken to be somehow above politics in the sense of cynicism and power play. Therefore, they were attributed the potential to act politically in a positive sense – not as participants, that is, but as outside observers: 'On the world, but not of the world', was a slogan of the movements.[43] An affinity, thus, with a view of politics as the application of science was influential from the early stage. It was also, however, fraught with tensions. They were connected to the control aspect of modern science – to its technical features, that is.

The fact that science is not only directed at facilitating human control of things but also lends itself readily to mechanisms of human control of humans, is of course disturbing to all who are committed to science as an expression of human freedom and democracy and as a means to achieve liberty. In some, the recognition of that tension resulted in futile attempts to purify science of its control aspects.[44] It seems, however, to have been more widespread to simply ignore the ambiguity and to commit to aims of getting rid of control and power mechanisms altogether and, at the same time, use science as a vehicle for the controlled liberation of humankind from hierarchies, power structures, inequality – and control.

In West Germany in 1967, the student activist Rudi Dutschke (1940–1979) declared his wish to realize the Garden of Eden on Earth. To succeed, he argued, and to achieve "liberation" from "power mechanisms", "revolutionary science" [*Wissenschaft*], "pervaded by politics" was needed.[45] Other visionaries were keen to revive dreams of using science to re-create and improve humankind and to create a whole new and perfected world.

The reinvention of political problems as wicked problems

Science expanded with ever new branches and this expansion was to some extent carried on or supported by the science enthusiasm exhibited by the student movements. But there was criticism as well. In 1973, reformulating in social science jargon what classical Aristotelian thought had characterized as practical (and, thus, political) problems, the term 'wicked problems' was coined by Horst Rittel (1930–1990) and Melvin Webber (1920–2006) as part of an ongoing discussion on planning.[46] They invented the term to denote problems, including 'nearly all public policy issues', that were unsuited to scientific problem solving because of disagreements about basic understandings, the unavoidability of making interpretations, the 'essential uniqueness' of each problem and the 'lacking opportunity for rigorous experimentation'. Actually, these kinds of problems could not be solved at all, Rittel and Webber claimed.

In line with classical political thought, but never referring to that fact, they made the case that such problems could only be re-solved – over and over again – by way of an 'argumentative process', 'critical argument' and 'incessant judgement' by participants. Furthermore, they added, problem understanding and problem resolution were 'concomitant to each other'. Thus, the very formulation of a wicked problem *was*, indeed, the problem.

Against that background, scientific engineering approaches in such areas as planning and management were denounced: '[T]he classical paradigm of science and engineering – the paradigm that has underlain modern professionalism – is not applicable to the problems of open social systems.' The argument was expanded to encompass the social professions in general: 'We shall want to suggest that the social professions were misled somewhere along the line into assuming they could be applied scientists – that they could solve problems in the ways scientists can solve their sort of problems. The error has been a serious one.'

When dealing with wicked problems, Rittel and Webber argued, the aim is not to find the truth but to improve some of the characteristics of the world people live in. Planners, they continued, were 'liable for the consequences of the actions they generate; the effects can matter a great deal to those people that are touched by those actions'.

The article has been used and quoted extensively. In particular, the notion of wicked problems has become widely used and may, whether by clumsiness or by intent, have contributed to increasing the already widespread animosity towards politics in the classical sense. Ironically, it seems to have only strengthened the general emphasis on social, as distinct from political, aspects of issues and to have spurred the further development and application of socio-technical methodologies as substitutes for practical-political approaches to practical-political problems.

Dialogue in vogue

The increased emphasis on social as distinct from practical-political aspects has been particularly notable in the area of communication. Dialogue has

Growth, Normality and Moneymaking

A problem was solved when, in 1986, genetic engineering facilitated the production of human growth hormone in microorganisms. Previously, human growth hormone had been in short supply because it had to be extracted from dead bodies. It was needed for a few very specific medical purposes such as the treatment of children affected by dwarfism. That problem was solved by the new technology, but another problem was created. The new solution – or rather, those who wished to produce it – were hungry for more problems.

In principle, all people who might benefit from growth hormone treatment because they suffered from a growth hormone deficit, could now have it. A huge market was in sight. Scientists, however, were unable to establish a standard for the normal production of growth hormone in healthy persons – but there was pressure from parents of children who seemed destined to end up as short adults. Three years after the technological breakthrough, Nordic paediatrics agreed on height standards. Adult men should reach at least 1.62 metres and adult women at least 1.50 metres to be regarded as normal. The standards prepared the ground for clinical tests of the effects of growth hormone treatment on healthy children who were estimated to become abnormally short adults. The treatment did not seem to make them taller, but the pressure from parents, in particular parents of short boys, did not go away.

Heightism made and makes itself felt and has affected exchanges – and probably decisions – about genetically engineered growth hormone all along. Eager producers are still around. And there is pressure on public purses. The issue, thus, encapsulates all the marks of a science-related public affair. Value judgements and social norms, motives of gain and technical-scientific aspects are intertwined. There are scientific and technological uncertainties. There is disagreement. True answers to the challenges presented by the technology and its movers cannot be identified. There is only the possibility of talking it over.

The textual snapshot about human growth hormone has drawn on Gitte Meyer, *Den kunstige krop* [The artifical body], 48–54.

been reinvented as a purely social concept, forming part of a cluster of related concepts – empowerment, inclusion and participation prominent among them – and linked to aims of socio-technical intervention to achieve positive social relations. Thus, rather than being connected to purposes of assessing the lay of the land, exchange in the sense of dialogue has come to signify the achievement of, for instance, intimacy, empathy and equality.

Primarily positive and optimistic assumptions about social relations between humans appear – for the time being, at least – to have gradually superseded primarily pessimistic versions. Accordingly, communication as dialogue seems currently to be connected to potential consensus rather than conflict, to inclusion rather than exclusion, to potential intimacy rather than separation and distance, to potential unity rather than disunity and to aims of promoting social equality rather than to attempts to uphold hierarchies. To a great extent, thus, dialogue has come to be viewed as a more likely means for the creation of social unity, consensus and harmony than abstract, universal truth. Perceived as a social cure-all, it has almost acquired a status as a competing truth – or as a means to fuse the poles of the truth versus the social dualism – in theories of reflexive modernity[47] and even in some idea(l)s of journalism.[48] Inclusion in dialogue and so-called participatory methods of a socio-technical vein have come to be seen as instrumental to the achievement of social equality, almost on a par with equal access to education.

Understandings of science communication have, of course, been affected by those developments.[49] Whereas the enquiry into and the understanding of reality have been maintained as the prerogatives of science, the view of communication as a means to bring humans closer to each other has, it seems, been accompanied by increased expectations of science as a potentially uniting force. In line with the early history of science, science communication has come to be seen by many as an enterprise aimed at gathering people around science as a shared cause.

The Classical Institution of Public Discussion

The classical institution of public discussion and deliberation was based on other premises and had other goals. First and foremost, it was founded on an understanding of politics as a substantial activity in its own right, tied to the notion of human life as always uncertain praxis, dependent in a variety of ways on the fact of the plurality and diversity of humans.

Exchange among points of view was seen not as a social remedy but as a way of shedding light on shared practical problems – matters relating to private households or economies were not considered public – that could neither be answered by religion nor solved by technical means. They were perceived, in other words, to be distinct, at the same time, from questions concerning universal truths and from technical problems, but might include elements relating to specialized knowledge.[50]

The framework, thus, relies on a distinction between different kinds of issues for enquiry, matched by corresponding distinctions between modes of enquiry. The view of human life as praxis and the identification of some problems as practical problems are accompanied by the concept of practical reasoning or phronesis, differing from technical rationality and from the contemplation of universal truth in much the same way that life as praxis is assumed to differ from the unlimited and the mechanical dimensions of reality.

Because of the use of force on objects that is inherent to technical procedures, classical thinkers, it has been argued, feared *techne* or technical reason as a possible threat to the freedom of public, political life.[51] As a temporal and personal kind of reasoning, assessing the lay of the land and the conditions for action from one case to another, phronesis does not possess such force. It has purposes, but no objects or aims of control. It is aimed at coping with the human conditions of limitations, uncertainty and diversity, not at getting rid of those conditions.

Although, thus, the logic of science and the political institution of public discussion were tailored to dealing with different problems – the scientific logic does not acknowledge the distinction, it should be kept in mind – some of their challenges are shared. Their responses to those challenges, however, are different. As a response to the fact that individuals cannot completely rely on their own senses and reasoning, that all make mistakes when we observe and assess specific situations, the scientific logic resorts to methodologies based on impersonal outside observation to identify true and correct answers. In contrast, the institution of public discussion – whether in antiquity or modernity – responds by insisting that multiple points of view should be taken into account and that there is no such thing as proof, only reasonable argumentation and critical assessments, when it comes to coping collectively with shared practical problems. It is founded on a principle of practical knowledge pluralism.

The institution makes sense only when based on those assumptions that connect the human condition with uncertainty and limitations *and* with the human capacities for thought and speech, facilitating the ability of citizens to deal with public affairs on the basis of co-responsibility. Speech has been described as the basic practical – and, thus, political – mode, connecting individual thought and different points of view. From this perspective, points of view – rather than individuals or groups – are being represented through speech. Communication or dialogue or debate or discussion are not viewed as instances of social relations but as shared enquiries. Such enquiries may reveal avenues of possible action, hitherto unthought of, while at the same time allowing different points of view to moderate and delimit each other.

Disagreement, thus, is not perceived as threatening but as a possible source of knowledge. In this context, the idea that for something to be genuinely public it must be above discussion,[52] tying benign public life to the ideal of consensus, makes no sense at all.

Originating in the Latin verb for shaking, to discuss – *dis-quatere* – literally signifies the activity of shaking something apart.[53] A discussion can be described as the activity of facilitating movement in a substance in order to properly examine it. Transferred to human affairs, it denotes a form of enquiry that draws on different perspectives on human reality.

As enquiries, discussions depend on critical audiences who are concerned with the substance of the issues that are up for discussion and do not allow themselves to be swayed by emotions into mere applause or rejection of speakers. The activity of enquiry is preconditioned by the existence of citizens who are both willing and able to participate critically and, thus, to pay thorough attention to the substance of speeches.[54] They depend on political equality and freedom of expression to make their contributions. And they need to concentrate on the issues. Thus, they should not be distracted by, for instance, concerns about social hierarchies insofar as they are irrelevant to the specific discussions.

In *The Fall of Public Man*, first published in 1977, British-American sociologist Richard Sennett spotted the above line of reasoning at work in the coffee houses of the early eighteenth century with their culture of exchange and debate. Characterizing the coffee houses as information centres, Sennett noted that as such they were naturally 'places in which speech flourished'. People went there 'to gain knowledge and information through talk'.

To enjoy oneself in a coffee house, Sennett went on, was

> a matter of talking to other people, and the talk was governed by a cardinal rule: in order for information to be as full as possible, distinctions of rank were temporarily suspended; anyone sitting in the coffeehouse had a right to talk to anyone else, to enter into any conversation, whether he knew the other people or, whether he was bidden to speak or not. It was bad form even to touch on the social origins of other persons when talking to them in the coffeehouse, because the free flow of talk might then be impeded.[55]

Coffee house speech, Sennett concluded – turning to the language of an outside observer of social relations – was 'the extreme case of an expression with a sign system of meaning divorced from – indeed, in defiance of – symbols of meaning like rank, origins, taste, all visibly at hand'.[56]

The idea(l) of public discussions as practical-political modes of enquiry has been maintained in some varieties of modern political thought but is absent in others. For instance, it has been observed that neither Thomas Hobbes nor John Locke (1632–1704), both prominent, early modern English philosophers, used a concept of public and political life that included public discussions and reasoning.[57] The concept of *Öffentlichkeit*, on the other hand – the German, Danish, Norwegian and Swedish terms are quite similar – is a descendant of the classical notion, but resists translation into English.

The standard translation into 'public sphere' is not only inadequate, but downright misleading. The spatial metaphor – public *sphere* – is not present in Öffentlichkeit.[58] It signifies neither a physical nor a virtual sphere or space but an activity or a series of activities aimed at opening issues to public scrutiny and exchange as part of the public formation of opinions.[59] It presupposes the existence of a critical public of political equals and can be seen as a residue of a tradition of enlightenment with an affinity for classical thought and with a fondness for critical discussions on aesthetic, literary and (meta)political issues. A close relationship between intellectual and political activity, perceived as activities of a kind rather than as opposites, has been prevalent within that tradition, which is now struggling to survive.

The activity of Öffentlichkeit makes sense only on the basis of a distinction between the practical and the technical, which has been losing ground for centuries. In 1972, German philosopher Hans-Georg Gadamer (1900–2002) noted that a technical understanding of practice as no more than the application of scientific knowledge had become standard.[60] The notions of the political and the practical had been more or less engulfed by those of the social and the technical, respectively. In effect, it was no longer possible to distinguish between technical and practical issues and to argue that they be treated differently.

At about the same time, another German philosopher, Jürgen Habermas – otherwise often at odds with Gadamer – made related observations. He saw the elimination of the distinction between the practical and the technical as the ideological core of a technocratic identity.[61] And, he added, in step with the marginalization from public exchanges of practical questions – and of the very notion of that category of questions – the public was also losing its function as a politically acting entity.[62]

The turn away from – or direct opposition to – the distinction between the practical and the technical, in short, is not a new phenomenon, and it is not a new insight that it is taking place. Actually, it has been observed; Thomas Hobbes rejected the distinction between praxis and production – as linked to technical activity and to aims of gaining control – already in the early seventeenth century.[63] The rejection, however, is likely to have been gaining ground ever since, following waves of science enthusiasm.

The application of a didactic science communication paradigm to all kinds of science-related topics is an example. The scheme does not take the different natures of technical-scientific and practical-political problems into account; it does not acknowledge the distinction. It makes sense only if all problems are technical problems, solvable by means of technical-scientific enquiries. If so, knowledge of the solutions would also be transportable. Only, that does not seem to be the case.

Technical-scientific enquiries have a beginning and an end; there is a moment of closure and after that the results may be packaged for transport. Whether seen as commodities or as common goods, they are supposed to be transferred from knowers to non-knowers, from scientific knowledge producers to knowledge consumers, from scientific specialists to lay citizen, the former disseminating their knowledge and teaching to the latter. In the context of doubt and disagreement, however – present, as a rule, in relation to science-related public affairs and political issues – the interpretational cluster of production and transport metaphors that has been setting the tone of discourses on science communication for decades becomes counterproductive. It was not developed to deal with the inexactness, uncertainties and multiple points of reasonable view that are peculiar to practical problems. Because it has no substantial idea of politics in its own right, it is more likely to hamper than to facilitate civilized discussions of science-related political issues.

Political Cultures in Nutshells: Traditions of Journalism

Academic and political cultures are intertwined. Drawing on shared historical experiences, varying from one place to another, understandings of science inform understandings of politics and vice versa. Science communication is one of those hot spots where understandings of science and politics come into direct contact.

To understand how science communication paradigms come about – and might be rethought – we need to understand how such interplays work out in practice. Different frameworks of thought about the mores and means of journalism are helpful to that purpose. They can be seen as models en miniature of wider frameworks or sets of ideas about politics, democracy and science, ideas that are linked to more basic assumptions about the natural order of things. Such frameworks, therefore, also provide a shortcut to understanding the background of different science communication idea(l)s. Here, we will look into two traditions of thought on journalism, tied, respectively, to a socio-technical and a practical view of politics. The former variety uses science as a model and shares some of its origins with science. The latter is more easily understood if one presupposes an affinity with classical political thought.[64]

The two traditions can be seen as representatives of Atlantic and northern European approaches respectively.[65] It is of particular relevance in a European context that they can also be seen as expressions of politico-cultural differences among language areas.

Because they have a history of conflict, the two approaches might easily be taken to constitute a dichotomy and, thus, to be mutually exclusive. They are, however, genuinely different, and although they have a history of mutual animosity and conflict – that, in turn, may be traced back to different enlightenment traditions – it seems more fertile to look at them as complementary idea(l)s of journalism than to presuppose a dichotomic relationship.

The reporter tradition

The Anglo-American reporter tradition has come to be widely seen as the epitome of modern journalism. The reporter is defined as a producer of news. Within this tradition of many schools, discussions about journalism tend to be framed as discussions about a social relation – which is also a social division – between reporters as producers and their audiences as users or consumers of news.

As a distinct framework of thought, the tradition evolved in the early twentieth century[66] in the wake of the American Civil War and seems to have been informed by the populist understandings of the public, public opinion and democracy that gained momentum in particular during the Progressive Era. Gradually, the tradition has come to include an ever-changing cluster of different schools – many of which might even be subdivided – such as those of investigative journalism,[67] public journalism[68] or precision journalism (advocating the use of social science methods in journalism). Representatives of different schools may be highly critical of each other, but when viewed from a certain distance their conflicts appear to take place within a shared logic that uses science as its yardstick and serves to define the reporter in the first place. Some favour a 'facts function', others a 'forum function' of the media, but the conviction that journalists ought to search for truth and unity is pivotal to all[69] and there is a shared affinity for standardized (impersonal) reporting recipes.

The notion of universal truth constitutes the centrepiece of the reporter framework and is tied to an ideal of unity.[70] However, because the notion of universal truth comes with a tendency to generate dichotomies, it also serves as a source of continuous ambivalence and conflict. Dichotomies, such as observation versus participation, facts versus feelings and the masses versus the elites abound. Not least, the tradition seems to be haunted by the assumption of a

The Politics of Happiness Science

Now and again, the happiest country in the world is appointed by representatives of the growing, multidisciplinary field of happiness research; described as happiness science by some. Within the field, drawing on methodologies from the exact sciences, a great many numbers are produced. The production of happiness figures and seeming facts do not, however, appear to result in consensus about the nature of human happiness and how it comes about. Instead, the numbers are subjected to wildly conflicting interpretations that serve as fodder for conventional political battles between left- and right-wing positions, or between proponents of Scandinavian welfare states and American market-based democracy.

The numbers show, some have it, that more should be done to equalize incomes, that people should be protected from working long hours and that the public welfare systems should be expanded with a particular view to preventing and treating mental disorders. Not at all, according to others. The numbers show, they argue, that people should be allowed the greatest possible freedom to improve their financial conditions and care for their families. Anarchically, the numbers have even been used as evidence that anarchy is the true way to human happiness.

The debate, thus, is a political debate. There is substantial disagreement. Posing as a scientific debate, however, norms for dealing with disagreement about inherently normative and evaluative concepts and issues are not in place. Science was not evolved in the first place to deal with such concepts and issues. Will the expansion of science in that direction alter some of its core features in the process? That is one of several challenges from the field of happiness research that science communication routines aimed at disseminating outcomes of scientific research lack the capacity to deal with.

The textual snapshot about the politics of happiness science has drawn on Gitte Meyer, *Lykkens kontrollanter: Trivselsmålinger og lykkeproduktion* [The happiness controllers: The measurement of well-being and the production of happiness].

fundamental dichotomy of truth versus the social: the reporter is committed to a notion of universal truth *and* is by definition a social creature, belonging to a social world of other such creatures.

In the aftermath of the American Civil War, the founders of the reporter tradition – drawing on experience of how journalism might stir 'sectional antipathies' and prepare the ground for violent conflicts[71] – seem to have opted for non-participation in human affairs as the safest bet, much like the founding fathers of the Royal Society looked for firm ground for modern science in the aftermath of the English civil wars. As the reporter tradition is marked by veneration of science[72] and uses it as its model, the generic reporter ethos was not of a participatory vein. Rather, participation, taken to be partisan by definition, was to be strictly avoided.

Generically, the reporter is defined as a producer of naked information to be transmitted to her social counterparts: the public as audience and consumers of journalistic products. The producer–consumer relation ties the reporter to her audience and disconnects her from it. She is supposed only to transmit naked observations of facts and events, and to consider herself an outside observer of human affairs, committed to professional values of non-participation and to meta-technical tasks of measuring and monitoring. News, as opposed to views, is valued. Information, the everyday, down-to-earth stand-in for knowledge, is valued. And conflict is a source of fearful fascination.

The framework evolved, it appears, as a response to conditions in societies marked by a fundamental belief in one single universal truth *and* by the perceived ever-present threat of violent conflict between groups who disagree on the nature of truth. Reporters, then, represent 'a longing for truth(s) beyond dispute' and they must provide 'the truth, beyond differences of opinion'.[73] Even though disagreement might be fascinating – perceived as 'disunity' and a possible source of catharsis[74] – it should be handled with care like other explosives. Responsible reporters, therefore, should always, in order to serve fair representations and to prevent the escalation of conflicts, include 'the opposite view'. The assumption that conflicts are characteristically two-sided appears to be based on – and is likely to reproduce – a monistic and dichotomic view of reality. The reporter inhabits a bipolar world where the professional values of objectivity, neutrality and impartiality are contrasted sharply with notions of advocacy, activism, bias, commentary, interpretation and partisanship. She strives to achieve purity – pure facts, pure news – untainted by the aforementioned contaminants.

As a producer of news, the reporter was (and is) not only presented with the task of getting the facts right (truth) but also with concerns regarding her social counterparts, the consumers and their presumed preferences and capacities.

Aims of inclusion were (and are) important, not only for commercial reasons but also to serve the democratic idea(l)s of social equality.

In the United States in 1959, the reporter tradition had reached maturity, social-scientific studies of consumers had gained momentum, and problems of compatibility between the reporter's two basic preoccupations – with truth and with social inclusion – had become apparent. This was described by the American publicist Douglass Cater (1923–1995) in his seminal book, *The Fourth Branch of Government*. Cater who was himself clearly committed to the reporter framework, for example, criticized the focus on the audience as frenetic consumers:

> News is big business. News is a commodity that must be purveyed to an ever expanding audience by increasingly monopolistic distributors. It must be homogenized for *Homo genus* in the mass. [...] There is the audience of his [the reporter's] readers, a frenetic group who, he is told, spend eighteen and one-half minutes a day reading five columns of news, of which only one-eighth is international. The reader, it has been said, is the median man, destined like Orphan Annie, never to grow an inch. [...] It is the median man's attention, not his intelligence that must be attracted and held.[75]

No stretching is needed for this to be connected to widely adopted journalistic criteria of sensation, dramatization and what's-in-it-for-me approaches or to note how far this takes the reporter from the ideal of merely transmitting naked observations. There is an obvious kinship with fascination as a highly ambiguous science communication idea(l), assuming that rational knowledge can only be disseminated to the masses of the people by way of appeals to irrationality: truth and reason are in trouble when they come into contact – as they do in this framework of thought – with their presumed opposites in the shape of mass audiences and politics.

The publizist tradition

The *publizist*[76] framework of thought on journalism appears to be – or to have been – thriving in particular, but not exclusively, in German-speaking and Scandinavian countries. Correspondingly, it has been placed within a north/central European media model. It represents a pluralistic approach to journalism and has probably been on the defensive since the rise of the reporter tradition and, even more so, since the end of the Second World War. Widely neglected in the field of journalism studies, it has a history of practice rather than of theory and empirical study. Characteristically, key notions of the tradition resist direct translation into English. The concept of Öffentlichkeit – the institution of public discussion – is just one of several possible examples.

Because of its pluralistic approach, the logic does not come with a tendency to form distinct schools – each advocating its own particular version of true journalism – and it may, at first glance, appear amorphous. Publizist journalists come in many different shapes and sizes and are of many different persuasions. What they share is a commitment to the institution of public discussion, which is ascribed the capacity to improve the understanding of real-life conditions for action and the ability of humans to actually act together in reasonable ways.

During the German Weimar Republic from 1919 to 1933, this approach to journalism was accompanied by a short-lived scholarly tradition of humanist journalism and newspaper studies.[77] German sociologist Max Weber (1864–1920) made a lasting contribution to the publizist framework of thought by linking politics to an ethics of responsibility for future action – as distinct from an ethics of ultimate ends, typical of religions – and describing journalism as the epitome of a political profession.[78] Half a century earlier, in France, science fiction writer Jules Verne (1828–1905) actually made a related observation in his long-unpublished 1863 novel, *Paris in the Twentieth Century*. The main character of the novel, bemoaning the disappearance of journalism, connected this to a sad decrease of interest in politics.[79]

For more than half a century both academic studies of and exchanges about journalism among journalists worldwide have used the reporter framework of thought and notions almost exclusively as their point of departure and reference. Thus, publizist norms have mostly been expressed in the shape of criticism – helpful to understanding the tradition – of features of mainstream journalism. Enquiry into publizist concepts that have remained in use constitute another possible means to a theoretical reconstruction of the tradition, even though such concepts may frequently have been reinterpreted to provide a better fit with the reporter tradition.

Rather than being defined, in the first place, as a producer, the publizist journalist is defined primarily as a participant or co-citizen – who is ascribed the specific task of editing (or *Redigieren*) an ongoing public discussion on public affairs. Journalism is considered an intellectual pursuit and includes reporting, independent analysis, interpretation and critique. In short, the journalist is expected to be critical in the classical sense of enquiring into issues from many different perspectives. She is ascribed the task of facilitating civilized exchange among different points of view. That, of course, includes the facilitation of the expression of disagreement, valued as a possible source of knowledge.[80]

Key notions of the tradition, such as *Aktualität* (topicality) and *Redigieren* (editing), originate in the Latin verb for action: *agere*. Whereas the reporter tradition's concept of news is habitually contrasted with views and, thus, with the exercise of judgement, the notion of Aktualität[81] is geared towards the identification of burning issues in need of public scrutiny and presupposes the exercise of judgement in journalists. They are not supposed to be outside observers or, for that matter, anonymous media workers.[82] Views, in short, are not perceived by definition as sources of suspicion, and the term is not used as a term of abuse. Interviews, as a consequence, may be conducted in the literal sense of meetings between views.

While 'to edit' comes from the Latin *dare* – to give; bring forth, produce[83] – the literal meaning of Redigieren is to bring or drive something back, to alter, compress or reduce something.[84] From this root, the activity has acquired different meanings in English, on the one hand, and in German and the Nordic languages on the other. Redaction in English means to remove information from a document because one does not want the public to see it.[85] It has come to denote a variety of censorship, mostly benign, but still censorship. In Danish, however – and this goes for other Nordic languages and German as well – *redigere* means to provide thoughts with an orderly form so that they may become fit for publication.[86] Rather than being considered a kind of bias, which ought to be avoided altogether, the activity of interpretation is seen to be pivotal to journalism. The need for properly exercised interpretation is emphasized and truthfulness is demanded. But the virtue of truthfulness is juxtaposed to lying[87] rather than – like (universal) truth – to falsity.

The style of reporting is not focused strictly on events but includes background and contextual information, not least of an historical nature,[88] serving to reveal the complexity of issues. Ascribing a capacity for thought and reasoning to the public, the perceived task is not to make it easy for people to make up their minds but to prompt reflection. To representatives of the reporter tradition this is one of several traits of publizist journalism that may be considered elitist.

Typical of the publizist mindset, German sociologist and publicist Siegfried Kracauer (1889–1966) published, in 1929, a sociological enquiry about white-collar workers as a series of newspaper articles, arguing that such enquiry made sense precisely to a purpose of stirring discussions in public.[89] Reporting in the sense of merely reproducing observations was, he found, insufficient to grasp reality. He used the metaphor of the mosaic[90] as a counter-image, symbolizing a preoccupation with the inclusion of a multitude of perspectives on reality and, thus, expressing the basic assumption that human reality is complex and multifaceted.

The publizist framework of journalism is vulnerable not only because of its lack of attention and recognition as a distinct journalistic framework of thought and practice but also because current forms and concentrations of media ownership encourage an understanding of journalism as a form of industrial production on a par with any other such production; the so-called production of scientific knowledge not excluded. Journalistic standardization and competition with a focus on entertainment are unlikely to maintain the kind of critical audiences that publizist journalism depends on. This development constitutes an existential threat to the tradition and can be seen as an indication that a process of marginalization of classical, practical thinking is still going on. The traits identified here – the commitment to exchange among different points of view, the emphasis on participation and on the exercise of judgement – all point to a kinship between publizist journalism and the classical notions of praxis and practical reason.

The reporter, the publizist and science communication[91]

The coexistence in Europe of such very different understandings of journalism, based on such very different understandings of politics and its possible relationships with science, illustrates the rich diversity of Europe or, wider, the Western world.[92]

This diversity also comes with a rich potential for mutual misunderstandings. One logic is at home in two-party systems, another in multiparty systems, with the former being much more open to religious rhetoric in politics than the latter. The differences between conservative and liberal attitudes in one system cannot be transferred to cover the differences between right and left in the other. And while the differences between conservative and liberal attitudes may move along anti- versus pro-science lines, stances towards science do not constitute a dividing line between North European right and left attitudes.

Of particular significance to our topic is the curious fact that the field of science communication, dominated by disseminative and didactic approaches, appears to have been somehow immune to the feature of diversity. Because the reporter and the publizist frameworks are rooted in different historical experiences, languages and conceptual understandings and are founded on different understandings of the nature of and interconnections among science, the public, and politics, it would seem reasonable to also expect them to result in different understandings of science journalism, mirroring understandings of science communication in a wider sense. This has not happened, however. As professional activities, science journalism and science communication have risen to prominence in the wake of the most recent wave of science enthusiasm and appear to have been influenced almost exclusively by understandings akin to those that shaped the Atlantic media model and the reporter tradition.

The fact that science functions as a model of the reporter tradition influences the idea(l) of science journalism and makes it incomparable with other areas of reporting. Science – as opposed to political opinions and power plays and to political or religious beliefs and zeal – is assumed to represent knowledge of reality, the search for truth and the highest standards of objectivity, neutrality and impartiality. Science, indeed, is above suspicion and the reporter is supposed to function as its servant. It is her task as a science reporter to disseminate scientific knowledge, inform public opinion, teach science to non-scientists and include them in the scientific endeavour. Scientists have performed the necessary enquiries and done the thinking. The reporter's task is merely to transmit the results of scientific efforts to a mass public of laypersons and to humbly reconcile herself with the deplorable fact that her transmission activities imply an unavoidable reduction of sophistication.

The assumption of a necessary reduction as a feature of science communication is linked to the assumption that politics is devoid of intellectual sophistication. This was distinctly expressed by American sociologist Robert K. Merton (1910–2003) more than seven decades ago. Addressing the challenge of taking scientific findings to policymakers, Merton argued: 'there is the problem of so formulating the findings that the most significant results will be intelligible to and engage the interest of the policy-maker. The "processing of the material" may require simplification to the point where some of the more complex though significant findings are discarded.'[93]

The publizist tradition of journalism, on the other hand, has been stagnating for at least half a century. Habits of publizist science-related journalism have not really evolved. Some basic approaches can, however, be deduced from the logic. Now and again they may even be encountered in practice.

Being of a pluralist vein, the publizist framework is not compatible with ideas of allowing any one institution a monopoly on reason and knowledge of reality. At the same time, however, and for the same reason, it is equally incompatible with any fundamental hostility towards science. It is not inclined, thus, to simply ignore scientific perspectives. The phronetic features of the framework imply that the question of how to go about the integration of scientific perspectives and findings into the wider societal context be assessed from one case to another, depending on the issue. The stress on public discussions among different points of view does not, of course, exclude scientists, and because participation in politics is not viewed as basically suspicious, scientists may openly, as authoritative voices, make a case without necessarily being labelled advocates or partisans. The logic, thus, has plenty of room for scientific knowledge claims and for scientists' assessments and arguments concerning science-related public affairs. But it collapses if those assessments and arguments are accompanied by demands for a straightforward deference aimed at halting discussions on topics that are not of a purely technical-scientific nature.

The science journalist is a science *journalist*. She is expected to be knowledgeable and competent within her field but so should journalists who practise in other fields. The journalistic task remains one of facilitating the public formation of opinions by way of civilized exchange among different, reasonable points of view. Journalists should enquire critically into different positions and make clear the vested interests of participants and stakeholders. That includes the positions and interests of scientists. Talking to scientists, she does not simply represent a lay and thus inferior position but another kind of reasoning than scientific rationality.

The role most suited to scientists within this framework is that of citizens with specialized knowledge or, if you like, of public intellectuals in a small way. To fill out that role, scientists must master the vernacular and possess knowledge about – not be ignorant of – the wider societal context of their speciality.[94] Most scientists, most of the time, simply wish to get on with their work, but now and again their specialized field of knowledge becomes topical. They should be ready, then, to contribute to public discussions – including exchanges with possible critics of their current projects – with knowledge claims, assessments and opinions.

The *science* reporter and the science *journalist*, no doubt, are very different creatures and have a lot to quarrel about insofar as they come into contact. They are, however, much too different to be opposites. They do not represent different normative valuations of shared assumptions but mirror science and politics (in the classical sense) as substantially different – and therefore complementary – activities. Thus, there is a potential for a division of labour and for mutual learning. It might be possible to adopt, to some extent, features from each other that might be particularly well suited to dealing with particular problems.

How to deal with science-related public affairs and political issues might be one such problem. In this case, the reporter logic is faced with the twin problems that it lacks ideas of politics as anything other than either the opposite or the application of science, and that scientific idea(l)s and assumptions form part of its foundation. It does not have the capacity to act as an interlocutor, providing other points of view vis-á-vis scientists who are concerned with these kinds of problem. The framework is bound to remain within the didactic science communication paradigm, which does not cater for discussions among different points of view insofar as science is somehow involved. Here, the publizist framework may have something to offer.

'Post-Truth': Prejudices about Politics Come True

The understandings of politics I have been discussing here are features of the history of modern science as much as they are features of the history of modern politics and democracy. Present challenges in the shape of science-related public affairs and political issues serve as reminders that the histories of modern science and modern politics are intertwined in much more subtle ways than the assumption of a science versus politics dualism indicates.

Modern science was born into an intellectual climate of religious fanaticism and despair of civic and political life. In some, this bred a kind of science enthusiasm that – continuing practices from religious strife – was expressed in polarized and polarizing forms of debate. These included the demonization of opponents and the zealous promotion of science as universal light. The enthusiasts, thereby, may have been generating – and may still be generating – the types of hostile opponents and the cynical, yet hot-headed political life they imagined in the first place.

More common among scientists, probably, has been a general wariness of politics; a lack of interest and a wish to avoid involvement, perhaps even a fear of contamination. That attitude, in turn, may not give much reason for concern as a trait in pure technicians. Such pure technicians, however, are rather scarce. More often than not, current scientific enterprises are related to political issues one way or another and may have bearings on how such issues are – or are not – resolved. The wariness, however, is still around. Only a few years ago, for instance, the BBC experienced difficulties when attempting to find contributors from science to debates about current affairs. Thus, it was reported in 2012, the team behind a specific programme had 'bid for many more potential panellists from the science world – but most refuse because they wish to talk about their field and do not want to become involved in current affairs'.[95]

To some extent, this wariness of politics in some scientists may be an indirect outcome of overenthusiasm in others. And to a certain extent, the presently much debated characterization of the present era as a 'post-truth' era – making widespread prejudices about politics come true – may be an outcome of the expansion to all areas of life of the scientific idea(l) of truth, rigidly opposed to the notion of opinions. It is a logical counterclaim to the idea that true answers and correct solutions can be found to all questions, no matter what, that no questions whatsoever can be answered that way and that 'my opinion' is all there is.

Shared by these two extreme all-or-nothing counter-positions are, first, their lack of distinction between (technical-)scientific and (practical-)political questions, second, their understanding of opinions as mere gut feelings and, third, their lack of distinction between the scientific notion of truth and the scientific and political virtue of truthfulness. Basically, they share the assumption of a science versus politics dualism devoid of any substantial ideas of politics as an activity in its own right. And both may, although by different routes, lead to the end of political democracy and its substitution by populism, technocracy or some hybrid of the two.

In 1962, American historian Richard Hofstadter feared sufficiently for the intellectual aspects of societal life in his country to write a book about American anti-intellectualism.[96] Only a few years later, in 1965, the British political theorist Bernard Crick published a defence of politics in the classical sense,[97] while the Jewish German-American political thinker Hannah Arendt continuously issued warnings against possible technocratic and totalitarian developments that could lead to 'the rule of neither law nor men but of anonymous offices or computers whose entirely depersonalized domination may turn out to be a greater threat to freedom and to that minimum of civility, without which no communal life is conceivable, than the most outrageous arbitrariness of past tyrannies has ever been'.[98]

At the time of writing, the warnings and appeals from such writers may have appeared oddly irrelevant. Why make anti-intellectualism an issue at a time of vivid intellectual exchanges? Why publish a defence of politics at a time when thousands and thousands of students seemed to be already – and very actively so – promoting politics as a cause? And, why warn against over-reliance on science and technology, when science critique was already the cry of the day?

Through the hindsight of today, it is easier to understand why. More or less unwittingly, the student movements of the 1960s and 70s revived and carried on the tradition of the anti-political movement of science enthusiasm, viewing science as universal light and striving to stretch the notion of scientific truth – and its technical equivalent: functionality – so that it might cover reality like a fitted carpet and govern all human affairs. The overall direction of the vivid intellectual debate was anti-intellectual. The seeming promotion of politics was based on understandings of politics as the opposite and/or the application of science. In general, the science critique was less concerned with the limits and limitations of science than with reforms that might facilitate its continued expansion. Meanwhile, critical attitudes towards the marketplace and its logic went hand in hand with the development of a strongly promotional culture, complete with techniques for achieving visibility and marketability.

The ways science is, and can be, spoken about today have been shaped by these developments. Giving a new lease of life to a reductive paradigm of science communication, closely related to reductive understandings of politics, they have left the current generations with multiple science communication challenges that relate to science as a societal institution and cannot be resolved within the framework of that reductive paradigm. Chapter 5 specifies and discusses some of these challenges and suggests the introduction of a political category of science communication as science discussion, suited to science-related political issues.

Notes

1   William Henry Smyth, *Technocracy: First, Second and Third Series* and *Social Universals*, pt. I, 6; pt. I, 15.

2   Ibid., pt. II, 8.

3   Ibid., pt. II, 8; pt. II, 13; pt I, 1.

4   Not least, a narrative about Danish culture as pervaded by trust – rather than as merely a culture not pervaded by suspicion – has been around for some decades. See for instance, Masamichi Sasaki and Robert M. Marsh, *Trust: Comparative Perspectives*.

5   Theodore M. Porter, *Trust in Numbers: The Pursuit of Objectivity in Science and Public Life*, 149.

6   Ibid., 122–23.

7   Gordon S. Wood, *The Radicalism of the American Revolution*, 245.

8   Richard Hofstadter, 'The Paranoid Style in American Politics'.

9   Bernard Crick, *The American Science of Politics: Its Origins and Conditions*, 43.

10   Edmund S. Morgan, *Inventing the People: The Rise of Popular Sovereignty in England and America*, 304.

11   Until relatively recently at least, the use of such terms as 'politics' and 'political' with almost exclusively negative connotations has been more common in English than in for instance German and the Nordic languages. For a discussion of this, see Gitte Meyer and Anker Brink Lund, 'International Language Monism and Homogenisation of Journalism'.

12   Irene Coltman, *Private Men and Public Causes: Philosophy and Politics in the English Civil War*, 148–51, 178.

13   Blair Worden, *The English Civil Wars, 1640–1660*, 161.

14   James Simpson, *Burning to Read: English Fundamentalism and Its Reformation Opponents*, 231.

15   Peer Sloterdijk, *Die Verachtung der Massen*.

16   Thomas Paine, *Common Sense*, 6–7.

17   Ibid., 5.

18   Coltman, *Private Men and Public Causes*, 188.

19   Gustave Le Bon, *The Crowd: A Study of the Popular Mind*, bk. I chap. 1.

20   Diana C. Mutz, *Hearing the Other Side: Deliberative versus Participatory Democracy*.

21   A. S. Hornby, ed., *Oxford Advanced Learner's Dictionary of Current English*.

22   As a political term, 'partisanship' cannot be directly translated into for instance German and the Nordic languages. In those languages, the term 'partisan' denotes

no more and no less than a participant in guerrilla warfare. See also Meyer and Lund, 'International Language Monism and Homogenisation of Journalism'.

23   The substitution of technical for practical goals in politics was observed by Jürgen Habermas, 'Technik und Wissenschaft als "Ideologie"?', 503–4.

24   Crick, *The American Science of Politics*, 11.

25   Robert A. Dahl, *A Preface to Democratic Theory*, 44.

26   Crick, *The American Science of Politics*, 54, 12.

27   The understanding that politics is defined by participation is still around. See for instance Volker Gerhardt, *Partizipation: Das Prinzip der Politik*.

28   The terminology of the public arena comes with connotations of struggle. An 'arena' is a sandy battleground. The term originates in the Roman Empire, as distinct from the classical Greek term for the public meeting place: the *agora*. Robert K. Barnhart, ed., *Dictionary of Etymology*; *Duden, Das Herkunftswörterbuch*. See also Markus Lang, 'Der Marktplatz: Ort der entpolitisierten Öffentlichkeit', 68–69.

29   Porter, *Trust in Numbers*, explores the perceived connections between numbers and democracy; Philip E. Converse, 'The Nature of Belief Systems in Mass Publics (1964)', 2, found that democratic theory 'greatly increases the weight accorded to numbers in the daily power calculus'; and Seymour Martin Lipset, *Political Man: The Social Bases of Politics*, 129, took the assumption of wisdom in the public vote for granted.

30   Slavko Splichal, 'Från opinionsstyrd demokrati til globala styrelseformer utan opinion' [From public opinion-based democracy to global forms of governance without public opinion-formation].

31   Whether in the 1870s, the early or late 1920s, the 1950s, the 1960s or 2005 – there seems always to have been a crisis of democracy, increasing corruption in politics and decreasing participation by citizens. Crick, *The American Science of Politics*, 32; Smyth, *Technocracy*; John Dewey, *The Public and Its Problems*; Fred S. Siebert, Theodore Peterson and Wilbur Schramm, *Four Theories of the Press*; *Port Huron Statement*; Demos, British Council, SNS and the UK Presidency of the EU, *The Network Effect: Connecting Europe's Next Generation Leaders: Media and Legitimacy in European Democracy*.

32   The suspicion of representative systems may have found particularly fertile ground in two-party systems where the winner takes all. George Orwell, *Politics and the English Language*, for one, seems to be indicating as much.

33   Aversion to middlemen can be seen as a Reformation legacy, transformed and multiplied many times over from its origin in protests against the Catholic priesthood's monopoly on the interpretation of the Bible.

34   Jean L. Cohen and Andrew Arato, *Civil Society and Political Theory*, for instance, used the term 'elite democracy'.

35   During the Progressive Era in the United States, '[t]he "Age of Science" and the "Age of Democracy" became commonly seen as all but identical concepts, the complementary progenitors of Progress', according to Crick, *The American Science of Politics*, 52.

36   John Desmond Bernal, *The Social Function of Science*, 404.

37   Nico Stehr, *Knowledge Societies*, 168, even referred to scientific knowledge as a kind of 'coercive power'.

38   Some see scientists as power holders – holders of knowledge as power, that is – on a par with other power holders. See for instance Wolfgang C. Goede, *Civil Journalism & Scientific Citizenship: Scientific Communication 'of the People, by the People and for the People'*.

39   The following quotations about sociocracy originate in Lester Frank Ward, 'Sociocracy', unless otherwise stated. Ward's italics.

40   Quoted in Leon Fink, *Progressive Intellectuals and the Dilemmas of Democratic Commitment*, 15.

41   Ibid., 24.

42   Tom Hayden and Dick Flacks, 'The Port Huron Statement at 40'; *Port Huron Statement*.

43   Ingrid Gilcher-Holtey, *Die 68er Bewegung: Deutschland, Westeuropa, USA*, 28.

44   Habermas, 'Technik und Wissenschaft als "Ideologie"?', noted the attempts to lib-erate science from its aspects of control and found that they constituted a blind alley.

45   Manfred W. Hentschel, 'Wir fordern die Enteignung Axel Springers'. Double quota-tion marks indicate that I have translated from the German.

46   Horst Rittel and Melvin Webber, 'Dilemmas in a General Theory of Planning'.

47   Anthony Giddens, *Beyond Left and Right: The Future of Radical Politics*, 19–20, 112–13.

48   Theodore L. Glasser, ed., *The Idea of Public Journalism*.

49   Even public authorities, such as the European Commisson, have adopted the slogans of participatory approaches and inclusion in dialogue in their science communication policies; see European Commission, *Public Engagement in Science*.

50   Aristoteles, *Retorik*.

51   Hannah Arendt, 'Kultur und Politik'.

52   Among numerous possible examples, see Porter, *Trust in Numbers*, 178.

53   Barnhart, *Dictionary of Etymology*.

54   Aristoteles, *Retorik*.

55   Richard Sennett, *The Fall of Public Man*, 81.

56   Ibid., 82.

57   Jürgen Habermas, *Borgerlig offentlighet: dens framvekst og forfall: henimot en teori om det borgerlige samfunn* [The structural transformation of the public sphere: An inquiry into a category of bourgeois society], 86.

58   Hans J. Kleinsteuber, 'Habermas and the Public Sphere: From a German to a European Perspective'.

59   As an activity, Öffentlichkeit does not conform to the standards of actual or imaginary things and it does not make a lot of sense to study it as such.

60   Hans-Georg Gadamer, *Truth and Method*, 556.

61   Habermas, 'Technik und Wissenschaft als "Ideologie"?', 513.

62   Ibid., 504.

63   Otfried Höffe, *Thomas Hobbes*.

64   The descriptions of and discussions about the two frameworks of thought on jour-nalism are much indebted to Gitte Meyer and Anker Brink Lund, 'Almost Lost in Translation: Tale of an Untold Tradition of Journalism'.

65   Daniel C. Hallin and Paolo Mancini, *Comparing Media Systems: Three Models of Media and Politics*.

66   Michael Schudson, *The Power of News*.

67   Dick van Eijk, ed., *Investigative Journalism in Europe*; John Pilger, *Hidden Agendas*.

68   Glasser, *The Idea of Public Journalism*.

69   David Paul Nord, *Communities of Journalism: A History of American Newspapers and Their Readers*.

70   For a discussion of the possible religious roots of the reporter tradition's commitment to truth – or Truth – see Doug Underwood, *From Yahweh to Yahoo! The Religious Roots of the Secular Press*.

71   Douglass Cater, *The Fourth Branch of Government*, 85.

72   J. Herbert Altschull, *From Milton to McLuhan: The Ideas Behind American Journalism*.

73   Géraldine Muhlmann, *A Political History of Journalism*, 17, 6.

74 Cater, *The Fourth Branch of Government*, 18.

75 Ibid., 171.

76 I use the German spelling of *publizist*, with a z, because *publicist* in English seems to be more frequently used as another term for an advertising or PR agent.

77 Stefanie Averbeck and Arnulf Kutsch, 'Thesen zur Geschichte der Zeitungs- und Publizistikwissenschaft 1900–1960'; Hans Bohrmann, 'Als der Krieg zu Ende war: Von der Zeitungswissenschaft zur Publizistik'; Hanno Hardt, 'Am Vergessen scheitern: Essay zur historischen Identität der Publizistikwissenschaft, 1945–68'.

78 Max Weber, *Politik als Beruf*, 36–37, 70.

79 Jules Verne, *The Lost Novel: Paris in the Twentieth Century*.

80 Hans Magnus Enzensberger, *Einzelheiten I: Bewusstseins-Industrie*; Eberhard Rathgeb, *Die engagierte Nation: Deutsche Debatten 1945–2005*.

81 Enzensberger, *Einzelheiten I*.

82 Ryszard Kapuściński, *Notizen eines Weltbürgers*, 108.

83 Barnhart, *Dictionary of Etymology*.

84 *Duden, Das Herkunftswörterbuch*.

85 Hornby, ed., *Oxford Advanced Learner's Dictionary of Current English*. Eighth edition.

86 *Ordbog over det danske sprog* [Dictionary of the Danish language].

87 Kapuściński, *Notizen eines Weltbürgers*, 108.

88 Ibid., 122, 276.

89 In Siegfried Kracauer's own phrasing: '*die öffentliche Diskussion aufzurühren*'. Kracauer, *Die Angestellten*, 8.

90 Ibid., 8, 15–16.

91 For an early version of this comparison, see Gitte Meyer, 'Encounters between Science Communication Idea(l)s: A Comparative Exploration of Two Science Communication Logics, with a Focus on Possible Conflicts and Potential for Mutual Learning'.

92 The fact that I have looked into two journalistic traditions should not be taken to imply that I take those two traditions to be the only ones existing in Europe.

93 Robert K. Merton, *Social Theory and Social Structure*, 278.

94 Russell Jacoby, *The Last Intellectuals: American Culture in the Age of Academe*.

95 BBC Trust, 'BBC Trust Review of Impartiality and Accuracy of the BBC's Coverage of Science: Follow Up, 12. I have come across similar attitudes when interviewing scientists.

96 Richard Hofstadter, *Anti-intellectualism in American Life*.

97 Bernard Crick, *In Defence of Politics*.

98 Hannah Arendt, *Sonning Prize acceptance speech 1975*.

Chapter 5

A POLITICAL CATEGORY OF SCIENCE COMMUNICATION

Sapere aude is an enlightenment motto of rich ambiguity, reminding us of the multiple and to some extent conflicting ideas of enlightenment that made the Enlightenment era so fertile. Because of the ambiguity, the motto is not easily translated from the Latin. Should it be translated into 'dare to know'? Or should it rather be translated into 'dare to make use of your own reason' or just 'dare to think'?

None of those very different translations – all in use – is false and none is universally correct. They are equally valid. Together, they mark an interpretational space. Daring us to know *and* to think, the space is useful to reflections and exchanges about how to communicate about science. When should predominantly didactic approaches, in the sense of 'dare to know', be used? When would dialectical approaches, in the sense of 'dare to think', be more suitable? There is tension between those understandings and approaches, but they are not opposed in a straightforward way. Rather, they are complementary. But we do have to think hard, from one case to another, to strike a proper balance.

Frequently, there is good reason to proceed along predominantly didactic lines, emphasizing the dissemination aspect of science communication, the sheer transportation of scientific knowledge from a group of knowers to others who lack and might benefit from that knowledge. But more and more often, knowledge claims concern huge and inexact societal questions, fraught with the kinds of uncertainty and complexity – including conflicts of interest – that are the hallmark of political issues proper and with ample room for different, reasonable assessments. Science-related public affairs tend to come with such features. Why not, then, proceed along predominantly dialectical lines in those cases?

The introduction of a political category of science communication, stressing the discussion aspect of science communication, would constitute a deviation from mainstream understandings of the mores and means of science communication. Some might even consider it a dangerous deviation and a threat to the authority of science. To science as an intellectual enterprise, however, it is hardly healthy to be granted such unlimited authority that no critique and sceptical questioning is permitted even when science transgresses the hard-to-distinguish borderline between the scientific domain of exact questions and the vast area of inexact questions.

As an intellectual enterprise, modern science is, at the same time, indispensable to and dependent upon modern democracies that carry on pluralistic discussions among different points of view, also concerning science-related public affairs and political issues. Such discussions may serve, among a great many other things, to delimit the area of scientific truth-seeking and problem solving and thus provide science, as a body of knowledge and rational methodology, with boundaries – within which scientific specialists can be acknowledged as authoritative voices – and with a context inhabited by possible interlocutors from other walks of society. Bodies without boundaries and context cease to be bodies. They explode or implode or just fade away and become unrecognizable.

Both as a body of knowledge and rational methodology and as an intellectual endeavour, science is more likely to be nurtured than harmed by the disagreements, contradictions, critiques and non-scientific perspectives that inevitably form part of public discussions on science-related public affairs. At the same time, the cultivation of such habits of discussion can be seen as a possibility for democratic knowledge societies to cope with the expansion of science in a reasonable way, steering clear of the pitfalls of populism and technocracy, allowing ordinary citizenship to scientists and integrating science as a societal institution proper.

Science Communication Challenges

Current and rather urgent science communication challenges relate to science in its capacity as a societal institution. More specifically, they relate to publicity seeking accompanied by, on the one hand, temptations to oversell the possible outcomes of research projects, and on the other hand, incentives to conceal or play down possible conflicts of interest or disagreements among scientists and to keep silent about aspects of uncertainty. They cannot be dealt with on the basis of deficit models of the public and related assumptions – often apparently exorcised but nevertheless alive and well in widespread science communication practices – of a radical science–society divide, placing scientists outside the sphere of social interests.

Hype and concealment

Sometimes scientific researchers promise too much. They oversell or hype their research, often probably with a little or a lot of help from professional communicators. They are hoping too loudly for technological breakthroughs. They continue a very long history of knowledge boasting.

More than two millennia ago, Aristotle pointed to boasting, and in particular to boasting motivated by self-interest and with a view to gain, as the worst of the vices corresponding to the virtue of truthfulness. The boaster, according to his definition, was one who pretended to have 'distinguished qualities which he possesses either not at all or to a lesser degree than he pretends'.[1] And those boasters whose object was gain claimed qualities that 'both convey some advantage to their neighbours and can escape detection as being non-existent – e.g. prophetic powers, or philosophical insight or medical skill'.[2]

Not only is the practice of knowledge boasting still with us as a regrettable feature of science communication, but incentives to practise it have also increased. We may not be dealing with a novel phenomenon, but certainly with a pressing one.

In 2001, an international group of researchers from the field of science studies drew attention to promises that were 'based upon a potential that is difficult to assess properly and which will take time to develop fully, but which are amplified through the media, excite the imagination of industry and the public and influence decisions about which parts of basic research are to be funded and which lines of inquiry are to be pursued'. The group referred to 'a thin line between authentic belief in the future potential and mere rhetoric of "selling" a particular line of research to politicians and the public'. Increasingly, it was argued, researchers adopted 'sales techniques when trying to obtain funding for what are in reality no more than options or potential spin-offs of unknowable research results'.[3]

An array of financial motives, including competition for funding, is among the incentives to oversell or hype[4] the potential outcomes of research projects, as is the aim of achieving legitimacy as potential problem solvers in a more general sense. Moreover, excessive enthusiasm among scientists – concerning science and, in particular, their own line of research[5] – is probably another forceful driver of hype. The latter variety may be particularly difficult to deal with, especially against a background of widespread and sincere belief that science is the epitome of reason and realism and represents a good cause in its own right as 'a limitless capacity to handle all that comes our way, no matter how complex and unanticipated'.[6] Within that sort of context, it may not be easy to digest the profane proposition that scientists, like other humans, may have a capacity for obsession.

Possible and relatively recent large-scale examples of hype in science communication might include the debate that took place in the 1960s about the expected human colonization of other planets, the debate in the 1990s about xenotransplantation or the even more recent debates on human cloning or the swine flu epidemic. However, the more insidious everyday variety of hints – advanced at random by scientific researchers – that results from a particular line of research may be ready for use in five to ten years time forms part of the overall picture. So does the propensity to dress up outcomes from research into highly normative issues such as human well-being and happiness, as if they were the outcomes of exact scientific enquiry into exact questions. An air of exactness, preparing the way for strong knowledge claims, is achieved by the extensive use of exact numbers, tables, graphics, exotic abbreviations and engine-like models. So forceful and persuasive are such modes of presentation that – although initially they may have been chosen simply to accommodate the mainstream or to appear convincing to potential funders of research – they may even serve to persuade the researchers themselves.[7]

The funding problems that function as drivers of overselling and hype may also work the other way around and result in the concealment of scientific findings or aspects of relevance to such findings. Confidentiality clauses may be included in contracts when scientists are contracted to do research for commercial companies or public authorities. Obviously, the clauses may result in scientists withholding or postponing the publication of information or assessments, and even the internal communication among scientists may be adversely affected.

The rights and wrongs of confidentiality clauses have been widely debated. There is no consensus. From one position – which may be the majority position among scientists – it is argued that, as a minimum, research that is carried out at public research institutions ought to be publicly accessible and thus exempt from demands for confidentiality. From another position the case is made that access to risk capital is preconditioned by confidentiality clauses. Submission to conditions of confidentiality, therefore, is seen to be also a precondition of scientific progress. Along related lines, the question of whether or not demands for confidentiality can be combined with independent research is subject to different assessments.[8]

Discrepancies between such positions point to a more basic disagreement – rarely discussed – about the understanding of the idea(l) that knowledge is, or should be, a common good. One interpretation has it that public access to scientific knowledge is the proper embodiment of the understanding of knowledge as a common good. Another interpretation, taking knowledge to be synonymous with scientific progress, accepts a degree of secrecy as a necessary means to furthering knowledge – as scientific progress – as a common good.

Demands for confidentiality are, at the same time, an obstacle to and a potential issue for science communication. As obstacles, directly at odds with aims of dissemination, they have not been ignored. The demand that sources of funding be disclosed is becoming standard in serious academic journals and serves, among other things, to direct attention to the conditions for the funding of science. As an issue for communication about science in a wider societal context much remains to be done.[9] It is a tension-loaded issue, complete with disagreement among scientists and the raising of questions about the conditions for carrying out scientific research. As such, it goes far beyond a framework of disseminating knowledge claims.

Uncertainty about uncertainty

A particular variety of concealment concerns aspects of uncertainty. Scientific uncertainty has become a key term in the science–society discourse and attempts have been made to find ways to deal communication-wise with this disturbing and apparently novel aspect of modern or postmodern science.[10] A narrative has evolved about so-called common people who – as opposed to scientists – are supposedly fearful of and unable to come to terms with uncertainty. The narrative may originate partly in vicarious motives, conveniently bypassing the fact that modern science evolved partly to make the world a safer place, has been driven all along by aims of achieving control of things and, thus, has never been comfortable with uncertainty.

As a scientist, assumed to be on top of things, it may not be easy to admit to being uneasy with uncertainty. A possible way of escape might be to project, almost as an act of exorcism, that quality onto others. As a European bioscientist once explained to me during an interview, he preferred not to refer directly to uncertainty and ambiguities when talking to others about his research: 'It would be discomforting and unconvincing, I guess. There should be a clear message. If you started getting mixed messages, support would evaporate rather quickly. As a society, we want quick, simple messages.'[11]

In many ways the application of scientific knowledge *has* actually made the world a safer place and has reduced human vulnerability to many natural onslaughts. Increasingly, however, it has been noted that new uncertainties, to some extent brought about by the very application of scientific knowledge, have taken the place of the uncertainties that have been brought under control.

One of the reasons why the classical notion of human life as praxis was discarded at an early stage of modernity may have been its insistence that life is uncertain and the consequences of human actions unpredictable. Early scientists set out to prove this wrong. Later, this attitude crystallized into the concept of progress.

Today's scientists are uncertain about how to deal with the persistent fact of uncertainty in science. If viewed from a classical, practical perspective, the fact is merely an expression of the basic human condition. As the use of scientific methods and approaches has expanded into evermore walks of life, scientific enquiry has come to be increasingly concerned with human affairs and practical, political issues. It was only to be expected, then, that increasingly the condition of uncertainty would make itself felt. Scientific practice is a human activity, subject to the practical conditions of limitations, uncertainty, unpredictability and human diversity. These are general features of human life as praxis, elements of those limitations that form part of the human condition. Uncertainty is not a technical problem that can be solved but an indication of basic conditions that should be recognized. That recognition, however, is blurred by the specificity of the mystifying notion of *scientific* uncertainty. It may be obscured also by the seeming factualization of uncertainty that takes place when uncertainties are presented without qualifications in the shape of risk calculations with an aura of exactness and certainty.

Most confusingly, disagreements among scientists are frequently depicted as instances of scientific uncertainty – as signs, that is, of immature scientific enquiry that has yet to find the true answers to controversial questions.[12] Scientists only disagree, it appears, because they are still looking for the true answers. Thus, they do not really disagree. They merely lack sufficient knowledge. Their apparent disagreement is a transient deficiency. It is also an embarrassing sign of weakness in the scientific community, preventing scientists within a more or less well-defined field from reaching a consensus and, thus, present a united front towards the outer, societal world. But do the publics of modern knowledge societies really crave a united scientific front? Are they unable to cope with the existence of disagreement? Or is that inability rather an historically conditioned feature of the logic of science?

Illusions about science and scientists do not constitute a stable ground for exchanges about science. It has become urgent to further the acknowledgement, among scientists and others, of science as a human enterprise that may help us cope with but is unable to escape the human condition of uncertainty – a condition that scientists are no less likely to be uncomfortable with than their fellow humans. Equally urgent is the recognition that substantial disagreement among scientists does occur and that this is only likely to increase as the use of methods from the exact sciences are expanded farther into areas of inexactness. Current developments, however, do not appear to be furthering that sort of acknowledgement.

Public opinion and scientific consensus

The notion of 'the scientific consensus' has come into wide use as an interim solution to instances of disagreement among scientists. As a term, 'consensus' indicates agreement, but the notion of the scientific consensus is only used in cases of disagreement. It is deployed as a means of guiding the general public towards those scientific voices that represent the current majority in a more or less clearly defined scientific area and away from 'Fringe Scientists'.[13]

The notion of the scientific consensus and the ways it is used indicates a connection to the widespread assumption of a knowledge versus opinions dichotomy, mirroring the assumed science versus politics dichotomy. People who disagree are of different opinions, but, within their specialities, scientists are supposed to be knowers – as opposed to having opinions. We are dealing here with an understanding of knowledge according to which the notion of disagreement about knowledge questions almost amounts to a contradiction in terms insofar as such disagreement cannot be reduced to methodological disagreement. Within this kind of logic, the very existence of a scientific community seems to be preconditioned by consensus and to be weakened by disagreement. Apparently, the event of disagreement brings science too close to the much despised area of opinions.

All these understandings were in use when, in 2011, the BBC Trust, as a key part of a review of its science coverage commissioned an emeritus professor of genetics to make an evaluation[14] that should 'include not just natural sciences but also coverage of technology, medicine and the environment relating to the work of scientists'.[15] The decision to initiate a review was triggered by controversies relating to the debate on climate change, but had much wider implications.

As a consequence of the review, the Trust decided to partly suspend the general demand that journalism should be balanced. Thus, the coverage of science-related issues should instead be guided by a principle of 'due impartiality' or 'due weight', linked to the notion of the scientific consensus. Agreeing with the reviewer, the Trust found that 'there should be no attempt to give equal weight to opinion and to evidence' and that a 'false balance [...] between well established fact and opinion must be avoided'.[16]

The purpose of the principle of 'due weight' was, it was emphasized, 'to achieve impartiality in science reporting, especially in areas of very intense debate and divided opinion, such as climate change'. The Trust pointed out certain difficulties: 'The broad principle of "due weight" is, of course, easily explicable, and in practice the centre of gravity in some subjects can be readily identified. But in a wide range of areas (for example, badger culling, stem cell research, genetically modified food or nuclear energy) it is harder to delineate where the scientific consensus might lie.'[17]

The examples mentioned in the quote are typical examples of science-related public affairs and political issues. They include exact questions, but basically concern inexact issues. The notion of scientific consensus seems displaced. Why not, for instance, speak about the majority opinion?

As already indicated, the notion of opinion is widely despised. And the notion of public opinion is even more despised. As a term, opinion originates in the Latin *opinari*: to think, judge, suppose.[18] Like knowledge, thus, it is connected to the activity of thought. It has, however, been and is still widely used to signify common and conventional ideas, bringing it close to the notion of *doxa* – unreflected judgements, carried out almost automatically – as opposed to the thoroughness of critical activity. The latter interpretation was obviously employed by French mathematician, philosopher and co-editor of the *Encyclopédie* Jean le Rond d'Alembert (1717–1783) when he characterized the exercise of critique as the opportunity to shake off 'the yokes of scholasticism, opinion, authority, in brief: prejudice and barbarism'.[19]

When, in 1781, the term 'public opinion' first entered an authoritative English dictionary, it exhibited much more positive connotations, referring rather to the public formation of opinions in the sense of well-considered judgements than to public opinion in the sense of *doxa*. About two centuries later, German philosopher Jürgen Habermas described such public opinion formation as a process by which opinions, articulated as arguments in public, were refined into judgements.[20]

The notion of *the* public opinion – in the singular and as distinct from public opinion formation – does not operate with the existence of individual and conflicting opinions. The public, according to the notion, is of one opinion. The public is a unity. The public is a mass. Some understandings of democracy have it that public opinion, in that sense, should rule. Critique of and warnings against such understandings have been issued for more than 2,500 years.

It has been argued that the public, political life of societies – *in casu* the city states of antiquity – were likely to be eroded by attempts to achieve complete unity.[21] It has been argued that humans are so different that it makes no sense to speak of Man: '[M]en, not Man, live on the earth and inhabit the world.'[22] Humans differ from each other and represent different opinions. As a consequence, public opinion may be used to signify the majority opinion, but there will always be some who disagree. Where commitment to the idea(l) of the public opinion prevails, the various dissenters from the majority opinion may not be allowed to, or dare not, speak out even though they might have valuable contributions to make. Used directly or indirectly to close down discussions, the notion of the public opinion, it has been argued, acquires tyrannical features.[23]

On top of that, according to the critics, it is possible to manipulate the public majority opinion and to direct it in this or that direction. Thus, the notion of the public opinion not only has a tyrannical potential that may lead to the oppression of dissenting minorities, but can also be used tyrannically to manipulate the majority.[24]

Touching the key concept of democracy – *demos*, the people, the inhabitants of an area – these critiques constitute a grave challenge to democracy and have been taken seriously as such by political thinkers. A principle of pluralism in public exchanges is one of the outcomes of their efforts. It is aimed at preventing that just one, easily manipulated majority opinion becomes completely dominant.

Connected as it is to the perception of the public as a mass public, the notion of the public opinion also comes with mass public connotations about drama, speed and highly strung emotions. The notion of the scientific consensus, on the other hand, calls forth completely different connotations of detached, calm and thorough enquiry. Nevertheless, it shares significant features with the notion of the public opinion. It refers to a majority opinion among scientists in a more or – frequently – less clearly defined area of scientific research.

The shaping of scientific majority opinions concerning complex societal issues takes place in the form of processes of opinion formation. This is made reassuringly clear by references to, for instance, 'the accumulation of collective opinion', 'accepted interpretation',[25] 'sufficient consensus'[26] and to aims of 'win[ning] over peer scientists'.[27] Only, at some stage during the processes of opinion formation – probably when a majority opinion appears to have evolved – they somehow cease to be connected to opinions. Instead, the outcome of the processes, the majority opinion, is promoted to the rank of the scientific consensus; opinion is transformed into scientific knowledge. As such, it is by convention protected from sceptical and critical enquiry by non-specialists and by specialists adhering to minority opinions alike. From this point on, the exercise of critique of the majority opinion comes to be perceived as an expression of hostility to science and may be labelled as the 'manufacture' of doubt and uncertainty.[28]

The notion of the scientific consensus encourages the understanding of science-related public affairs and political issues as scientific rather than political issues. As a consequence, the authority of science is granted precedence and the room for exchanges among different points of view is diminished. This policy, however, appears misguided. Infused with tendencies to polarize and demonize, one of its possible side effects may easily be the generation of extreme counter-positions – the production, so to speak, of enemies to be fought down. On the other hand, references to a majority opinion among scientists, achieved by a process of reasonable opinion formation, would be less likely to have such effects. Paying tribute to the political character of the issues, it might contribute to another process of reasonable opinion formation in public.

The upholding of a we – they relationship between scientists (insiders) and other citizens (outsiders) – tied to a deficit model of the general public – is yet another problem relating to the notion of the scientific consensus. The discourse presupposes the existence of an almost existential gap between scientists and other citizens. The presumed counterparts of representatives of the scientific consensus – 'average citizens' and 'ordinary people' – are not, it seems, ascribed intellectual capacities that would allow them to be persuaded by the argument behind the majority opinion among scientists speaking as authoritative voices in a particular field. Instead, the public is given the stronger medicine of the scientific consensus, drawing on the authority of the exact sciences even when key elements of the issues in question are of an inexact nature. But that might, in the long-term, erode the general ability of democratic knowledge societies, pervaded by scientific enquiries and knowledge claims, to deal with the outcomes of such enquiries and to assess such claims.

Awe, banalization, imitation, quackery and superstition

The idea(l) of basic or pure science is – or was – among other things an idea(l) of scientific practice as a kind of activity that has no customers, no clientele. Sociologist Robert K. Merton (1910–2003) spelled this out about seven decades ago. The scientist, he argued,

> does not stand vis-à-vis a lay clientele in the same fashion as do the physician and lawyer, for example. The possibility of exploiting the credulity, ignorance and dependence of the layman is thus considerably reduced. Fraud, chicane and irresponsible claims (quackery) are even less likely than among the 'service' professions. To the extent that the scientist-layman relation does become paramount, there develop incentives for evading the mores of science. The abuse of expert authority and the creation of pseudo-sciences are called into play when the structure of control exercised by qualified compeers is rendered ineffectual.[29]

Merton drew the conclusion that '[t]he social stability of science can be ensured only if adequate defences are set up against changes imposed from outside the scientific fraternity itself'.[30]

Science has moved on since then. In today's knowledge societies, scientific researchers stand, and are expected to stand, vis-à-vis potential customers, clients and financial supporters all the time. They are present in the marketplace. They serve as specialist policy advisors, technology developers, problem solvers and suppliers of definitions. Accordingly, the idea(l) of pure science has lost ground. All science is now generally perceived to be at least potentially applied science. But what about the pseudo-sciences, fraud, quackery and credulity that Merton feared – and perhaps even spotted on the horizon – and could see no other remedies for than the radical isolation of the world of science from the rest of society?

If ever such isolation was a remedy for anything, by now, surely, it has become obsolete. Other remedies must be found to protect science from its vulnerabilities.

The vulnerability to quackery and to the development of pseudo-sciences originates partly in the fact that science, as a consequence of the prevailing stress on methods, is easily imitated. Activities have come to be widely regarded as scientific if they employ methods from or akin to those that are used by the exact sciences. Even the use of rhetoric from the exact sciences – the language of exact numbers and mysterious abbreviations – may suffice to provide claims with the guise of scientific evidence. To quacks, whether concerned with complicated and controversial political issues or with mere everyday trivia and banalities, this is good news. To others, it is a challenge to be tackled. In a worst-case scenario, all sorts of activity, irrespective of the subject matter, may resort to the seeming application of methods from the exact sciences and then proceed, drawing on the authority of science, to make strong knowledge claims and to practise disseminative science communication.

References to scientific methodology have acquired the force of a magical formula or spell – the ability, in other words, to fascinate. This is neither a recent trend nor a novel critique. The point that science has been turned into a 'fetish'[31] and that there is science credulity around 'to the point of superstition'[32] has been made over and over again. However, the propensity to regard science as a belief system seems to have increased alongside the expansion of science. That expansion, in turn, has enlarged the terrain where quacks, mountebanks and charlatans may successfully abuse the authority of science.

'We are being treated as the oracles we think we are,' an interviewee from the field of economy told me in 2002.[33] Claims are being made that 'science has already resolved questions that are inherently beyond its ability to answer' complained biologist Austin L. Hughes in 2012. He felt unable to connect the quality of modesty, that had attracted him to science in the first place, with the 'aura of hero-worship accorded to science and scientists'. And he pointed to a need for science to be protected from 'its potential for excess and self-devotion' unless a 'priestly caste demanding adulation and required to answer to no one but itself' be created.[34]

Hughes' critique concerned a perceived rise of 'scientism' – synonymous with science superstition. But superstition was one of the foes that modern science was expressly developed to counteract. Can there be such a thing as science superstition?

Dictionaries define superstition as 'a false or irrational religious belief or practice';[35] as distorted beliefs[36] or as beliefs in mystical or supernatural or extrasensory forces that, according to the dominant religious beliefs or views of nature, are exaggerated and unreasonable and may originate in fear or ignorance.[37] The term comes from a Latin term for '*excessive* fear of the gods, unreasoning religious belief or *awe*, perhaps originally meaning a state of religious *exaltation*'.[38] According to those interpretations, science superstition signifies exaggerated and unreasonable beliefs in or awe of science.

Ironically, such beliefs may be tied to understandings of science as the antidote to superstition. The movement of science enthusiasm was founded, it seems, on the latter belief, which has also, I have suggested, informed the dominant understandings of science communication. Fearing that superstitious beliefs might be generated if the authority of science as the epitome of reason and realism decreased, the maintenance of scientific authority in an almost authoritarian sense has been viewed as imperative.

The wish to liberate humankind from superstition may, however, also serve as a starting point for other lines of reasoning. It can be argued – I actually argue – that superstition is given free rein if continuous, critical discussion about ways of knowing and forms of knowledge is neglected. According to this line of reasoning, continuous critical discussions are imperative to the maintenance of science as an intellectual enterprise. And the generation of superstition is the likely outcome if science is ascribed the qualities of a belief system. Awe of science prepares the ground for, at the same time, excessive beliefs in science as a force for good and excessive beliefs in science as a force for evil. On top of that, the assertion of the authority of science in the strong, authoritarian sense may, in the long-term, undermine scientists' possibilities for gaining confidence and being recognized as credible and trustworthy authoritative voices.

As science penetrates further and further into societal practice, it has been argued, it can fulfil its societal function only 'when it acknowledges its own limits and the conditions placed on its freedom to maneuver'.[39] Present discourses on science-related public affairs, however, display other inclinations. Whether in political decision making, in speculations about future scientific and technological developments or in reflections on the ethical implications of such developments, scientific knowledge claims, including claims about future developments, are generally perceived as firm ground that can be used directly as points of departure and that only scientific peers – increasingly difficult to define – are allowed to question. Concurrently, methods from the exact sciences are applied to broad, complex and inexact topics or problems. Subject-wise, the research belongs in the humanities. Its aims, however, the methodologies and, not least, the resulting knowledge claims and rhetoric generally originate in the science tradition. As a net result, research efforts concerning broad and multifaceted topics with significant normative components are connected to aims of causal explanation. The conclusions are presented in the shape of strong knowledge claims. And policy proposals appear as the outcomes of 'sound science', disconnected from human assumptions and judgements.

Science, it appears, has no limits that might be worthwhile or even urgent to consider.[40] The different qualities of research topics – their exactness or inexactness – are of no consequence. Distinctions remain hidden beneath the surface of the currently expanding and seemingly moderate terminology of 'research'. In practice, the outcomes of research efforts gain acceptance as scientific knowledge almost automatically insofar as scientific methodology is applied. Life is made easy. Outside science – or 'research' – no thinking is needed. We have come close to realizing the bizarre development that Scottish philosopher Adam Ferguson (1723–1816) pondered two-and-a-half centuries ago in his treatise on civil society: '[T]hinking itself, in this age of separations, may become a peculiar craft.'[41]

The reintroduction of distinctions between research projects according to their topic might serve to make room in public exchanges on science-related public affairs for the activity of thought, crucial as it is to science as an intellectual enterprise. The making of such distinctions relies on the activity of thought and thorough appraisals from one case to another. Allowing reflections of that kind to form part of public exchanges, therefore, might decrease the propensity to almost automatically recognize as scientific – on the basis of the criterion of methodology – all knowledge claims that present themselves as scientific. To serve that purpose, however, the Mertonian demand for defences against the world outside the scientific fraternity would have to be put aside as directly counterproductive and more suited to strengthening than to dismantling widespread beliefs in the magic of scientific methodologies and the self-sufficiency of science.

Barriers to critical self-examination

There are many reasons why science needs interlocutors from other parts of society. One such reason is that the science tradition has a built-in impediment to confront precisely the kind of issues that it, as a human activity, desperately needs to come to terms with: its possible limits and limitations; its ambiguities, tensions and schisms; its basic assumptions, conceptual understandings, interpretations and aversions. By marginalizing such topics as non-scientific and inexact – which they are – or as highbrow, elitist or mystical, science trips itself up.

One of the fundamental tensions of the science tradition is constituted by, on the one hand, the idea(l) that science is and should be an outsider to society at large and, on the other hand, the ambition to effectively govern society by way of science-based interventions and policies. In recent years, much effort has been put into exorcizing the assumption that places science outside the social sphere. In practice, however, the assumption has remained effective.

The dominant logic of science communication as a didactic enterprise only makes sense on the assumption of an inside (science) versus outside (society) divide. It does not facilitate that science, practising reasoning and scepticism, is itself made the object of reasoning and scepticism – as a human activity among other such activities. The exercise of scepticism, in particular, has remained a scientific prerogative. The ever-increasing use of scientific methods has not been followed by a corresponding extension of the ethos and norms of science. But the methods and the norms did not evolve as completely separate entities. Does not the widespread application of the methods, but not the norms, make the application of the methods less valuable? And might not that kind of scientific practice backfire and, as a result, erode the ethos for good?

There seems to be a need for a reinterpretation of traditional, scientific norms to respond to current conditions.[42] Merton's codification of the scientific ethos, the CUDOS norms, provides a useful point of departure because each of the norms – Communism, Universalism, Disinterestedness and Organized Scepticism – can easily be understood, at the same time, in a descriptive *and* in a prescriptive sense. Whereas the former understanding relates to outside observations of how things seem to be, the latter understanding connects to ideals about how scientists ought to act. This is a distinct feature of Merton's codification, which is not replicated in, for instance, the more recent acronym PLACE, referring to Proprietary, Local, Authoritarian, Commissioned and Expert.[43] Aimed at capturing characteristics of industrial and/or post-industrial science, it was clearly launched as an updated substitute for Merton's codification but lacks the CUDOS qualities. Reluctant to deal with normative aspects, it does not facilitate ethical reflection but merely chases normativity underground where it might turn moralistic. Thus, I will hold on to the Mertonian codification.

The norm of organized scepticism, in particular, might be rethought as a norm of relevance not only to internal exchanges within individual scientific disciplines but also to exchanges across disciplines and to public discussions on science-related political issues.

Insofar as methods from the exact sciences are applied to complex societal problems – that are anything but exact – it would seem wise to introduce a political category of science communication that would allow and facilitate the exercise of reasonable critique and sceptical questioning in public exchanges on science-related public affairs. While improving the opportunities for the public at large to realistically evaluate the possible uses and limitations of scientific and technological projects – and, in Merton's phrasing, to distinguish 'spurious from genuine claims' to scientific authority[44] – it might at the same time highlight the fundamental uncertainty of science as a quality that does not make science less valuable, only more complicated to deal with in practice.

Admittedly, that kind of extension would conflict with cultural beliefs – more deeply rooted in some cultures than in others – and it would come with greater demands on all participants in exchanges on science-related issues than the dissemination paradigm. Scientists, for instance, would have to consider and concern themselves with the wider context of their topics and specialities. And ideas of science as the one and only model, not only of reason and realism, but also of civilized exchanges, would have to be modified.

A much debated statement by climate scientist Stephen Schneider (1945–2010) may serve to illustrate the latter idea. In 1989, Schneider[45] made the case that scientists were faced with the challenge of finding the right balance 'between being effective and being honest'. Whereas 'being honest' was linked to science, 'being effective' was linked to public and political life. As scientists, Schneider argued, 'we are ethically bound to the scientific method, in effect promising to tell the truth, the whole truth, and nothing but – which means that we must include all the doubts, the caveats, the ifs, ands, and buts'. However, in order to get messages through to the public, he found, scientists – like other people – needed to 'get some broadbased support, to capture the public's imagination. That, of course, entails getting loads of media coverage. So we have to offer up scary scenarios, make simplified, dramatic statements, and make little mention of any doubts we might have.'

A dispute linked to the statement, related to Schneider having been misquoted as a proponent of stretching the truth. The assumptions about the natural cynicism of political life that he took for granted in his statement seem, however, not to have been disputed. Most probably, they were generally taken to be commonplace. But that, indeed, constitutes a science communication challenge of some enormity. The widespread adoption of the understanding of public and political life as naturally cynical may justify – and, indeed, inspire – cynical behaviour in public, not least if there is a purpose of furthering science as a good cause in its own right.

Clearly, prior to being crystallized into different idea(l)s of science communication, the challenges and positions I have been discussing here concern different idea(l)s of science and, in particular, of the roles that science and scientists can and should occupy in society. Scientists' understandings, often implicit, of their own position in wider society, influence their approaches to their research and to their communication practices, and vice versa. Do they carry out research *on* human objects or *with* human agents?[46] And do they communicate their outcomes *to* lay audiences or enter into exchanges about the research *with* other citizens?

Understandings of science and of science communication are intertwined. And understandings of science form a much neglected science communication topic, falling outside the scope of mainstream understandings of science communication. Rather unfairly, these understandings of science communication leave scientists with either the whole responsibility for the quality of science communication, because they are the knowers, or with no such responsibility, because communication is not their speciality. With its focus on the dissemination of knowledge claims and its non-normative pretensions, the dominant science communication paradigm hampers much needed reflection and exchange – across national cultures, academic disciplines and research areas and between academics and practitioners from various fields – on how to deal responsibly with inherently normative and evaluative issues, topics and concepts in scientific research. Those problems are emblematic of today's knowledge societies. It does not seem right to leave scientists alone with them. But a wider understanding of science communication is needed to enable us to confront them.

A Possible Exit from the Elitism–Populism Axis

Not only classical political thought but also classical rhetoric along Aristotelian lines offer possibilities to rethink some of those basic assumptions that may distort science communication, are at odds with modern science as an intellectual enterprise and threaten to extinguish the pluralistic elements of its heritage. Prominent among these assumptions is the view of society as divided into the (emotional and non-intellectual) masses and the (intellectual and political) elites. That assumption, in turn, forms the basis of the idea – shared by elitist and populists – that the general public is almost exclusively susceptible only to appeals to emotions or self-interest. In contrast, the classical understanding of humans, because it took the capacity for thought and reasoning to be a general human feature, presupposed that political speeches were aimed at calling forth that capacity in audiences, and assumptions that the citizenry at large are unable to cope with uncertainty and disagreement have no place in that framework of thought.

Aristotle distinguished between *theoros* and *crites* audiences.[47] The former corresponded substantially to modern understandings of mass audiences, defined by having a narrow horizon, no intellectual leanings and no sense of wider responsibility. It was not, however, taken to constitute a social group or a composite of such groups. The distinction was made to clarify the practical problem that speakers might, depending on the appeals they made, generate a theoros audience – which would be disinclined to critically appraise the substance of a political speech. Appealing to that sort of audience, speakers – depending on how adept they were – might succeed in calling forth strong emotional responses, pro or contra a specific position, but no more.

To achieve a thorough and critical examination of public affairs – and that was taken to be the purpose of political speeches – political speakers were well advised to actually address a crites audience by appealing to the capacity for critical thought and reasoning. This was considered crucial to the function of the political institution of public discussion as a practical kind of enquiry into public affairs of a practical-political nature – concerning, that is in current usage, questions that could neither be answered by science nor by religion. The institution of public discussion could only fulfil its function if the exchanges were sufficiently thorough to allow the experiences, impressions and reasonings of individual participants to modify each other by means of a collective process of thinking aloud.

Having swallowed the basic assumptions, this line of reasoning is heartbreakingly simple to follow. As a rule, however, contemporary communication studies work on other assumptions. First and foremost, it is widely taken for granted that communication, prior to it being anything else, is a matter of social relations. As such, communication comes to be seen as a means to potentially maintain hierarchies, or to help speakers increase their social status, or to achieve intimacy and equality by way of a display of inclusive attitudes towards audiences.

Combining the social perspective with technical approaches, mass communication tends to be seen as a series of processes that produce social relations and, at the same time, transport messages from producers to consumers. Language serves as a means of transportation rather than as a medium for thought. Processes of communication that are based on social-scientific knowledge of the intended receivers of messages are generally preferred and may be supported by standardized guidelines, complete with socio-techniques and toolboxes. Processes of communication that are dominated by the views of the senders of messages, and appear to be indifferent to the audiences, are frowned upon as ineffective, arrogant, exclusive, elitist and undemocratic. Practical understandings of communication as processes of enquiry, serving to throw light on shared, practical issues and drawing upon the capacity for reasoning among the participants, are not included in the framework and are likely to be viewed as elitist.

To science as an intellectual activity that is widely hailed as the epitome of reason, however, it would not appear very far-fetched to rely on appeals to reason in its communication practices. Moreover, the distinction between technical-scientific and practical-political questions might be useful in more than one way to science in that capacity.

First, it would facilitate reflections and deliberations on the proper approach to science communication from one case to another. Second, it might further the recognition that although audiences may have a lack of technical knowledge of a certain topic, they may nevertheless be capable of following and contributing to arguments of a practical-political nature and to cope with the existence of uncertainty, disagreement and conflicting interests. This would not rule out the possible use of social-scientific knowledge to support communicative practices, keeping in mind that such knowledge may be tied in subtle ways to social prejudices and, thus, may serve inadvertently to uphold the hierarchies it was meant to dismantle.

As a complement to rather than a substitute for the social (and technical) perspective, the practical (and political) perspective might provide science communication with an escape route from the elitism–populism axis with its condescending assumptions about the public at large. That, in turn, seems necessary if serious attention is to be paid, outside scientific disciplines, to the substance of science-related public affairs and political issues.

Science communication as practical reasoning and scientists as citizens

From a practical point of view it is not a given that science- and technology-related issues shall be seen and debated exclusively or predominantly as scientific issues. Some such issues may be regarded as political issues with scientific elements, best suited to practical reasoning. The idea of science as universal light and problem solver is replaced, then, by a framework of practical knowledge pluralism, using public discussions among different points of view as a form of enquiry and dependent on critical audiences in the literal sense of audiences that explore and perform judgements about the substance and context of issues. The task of science communication becomes one of integrating, from one case to another, scientific elements into a wider and more complex societal context[48] and of introducing the issues into public discussions. The task of scientists becomes that of participating as citizens equipped with specialized knowledge.

As will be remembered, practical reasoning in the classical sense of *phronetic* reasoning is concerned, at the same time, with assessments of the lay of the land and the possibilities for fair and reasonable action. There is no assumption of a dualism of the purely normative versus the purely factual; actually, human affairs are assumed never to be pure in that sense. Consequently, there is no radical separation of ethical issues from knowledge questions, but ethical aspects are kept open as topics for thought, assessments and exchanges. Science-related public affairs can, in other words, be discussed as ethical-cum-knowledge questions, which is an advantage considering that one of the currently most urgent science-related ethical challenges – the challenge of coping with uncertainties relating to scientific and technological developments – is closely tied to knowledge questions.

The logic would fall apart if technical-scientific rationality were granted a monopoly on all kinds of knowledge of reality, leaving practical reason with the task of dealing only with purely normative or moral questions. Thus halved – into instrumentalism and moralism – it would lose its sense of reality and cease to be practical.

There is a place for scientific rationality and specialized, scientific knowledge within the wider framework of practical knowledge pluralism and reasoning, but the place comes, as places do, with boundaries. Because the boundaries do not follow the lines of a dichotomy of facts versus values and cannot be defined once and for all, they need continuous attention. This is a practical challenge in its own right and may give rise to continuous debate, accompanying and informing discussions of individual cases.

Suited to some but not all science-related questions, a science communication paradigm along those lines, dialectical in the classical Aristotelian sense[49] and based on a down-to-earth pragmatic appreciation of science as a human enterprise, would neither see science as a possible substitute for politics nor its opposite. Rather, scientific arguments, representing a particular perspective on reality, would be seen as necessary contributions to be taken into account in practical-political exchanges on science-related political issues.[50]

There is no denying that phronetic and dialectical approaches are more demanding to practise than approaches to science communication as a socio-technical activity of didactics. Practical reasoning is demanding. It is not aimed at making life easy but presupposes that life is difficult. It cannot be converted into techniques but emphasizes the virtue of good judgement in reasoners. If seen from a social perspective, that kind of emphasis might well be considered elitist. From a practical point of view, critiques along such lines should neither be rejected nor accepted at face value but be attended to in the general discussion on the roles of science and scientists in society.

The recognition of science as a societal institution proper, and of scientists as co-responsible citizens, would relieve scientists from the obligation to adopt the role of oracles. It would, on the other hand, oblige them to cultivate their ability to exercise critical openness towards arguments and points of view of a non-scientific nature.[51] Thus, it would be preconditioned by the recognition that reason and sound judgement may be found outside the province of science, and that science and scientists might gain from having to consider other perspectives.

Western disagreements and their possible global uses

Different political and academic traditions in Europe are connected to different language areas that harbour and accommodate different understandings of universally shared key concepts, notions and core values, including basic assumptions about knowledge and idea(l)s about the roles of science and scientists in society. Theoretically, the actual diversity provides ideal conditions for the development of different understandings of science communication and for a general openness to a variety of approaches. Reality, however, does not seem to work that way.

It has been argued that the 'kinds of explanations that take hold in a society reflect cultural beliefs'[52] and a wide-ranging comparison of encounters with biotechnology in the United States, Britain and Germany[53] has demonstrated how different political cultures respond differently, by way of different decision-making processes and choices, to possibilities that are or appear to be offered by the life sciences.[54] The same might apply to science communication logics. The kind of science communication idea(l) that takes hold in a society is likely to reflect cultural beliefs. In principle, different political and academic cultures foster different science communication logics. In practice, however, cultural export–import activities may prevent that from happening.

Serving as unspoken premises, assumptions about science and scientists may, it has also been argued, co-shape, in unpredictable ways, those encounters with actual science that, in turn, trigger different political cultures into reshaping themselves.[55] Along somewhat related lines, the unreflected import of a specific science communication paradigm may trigger the importing political and academic cultures into reshaping themselves to the best of their ability – which may not be very great – to conform to the exporting and dominant culture. The almost unisonous adoption in Europe of understandings of science communication as a didactic enterprise, aimed at disseminating and promoting science, is an example.

There are, as we have seen, other routes on offer. The rich variety of European political and academic cultures, already present in the multiple enlightenment versions of the seventeenth and eighteenth centuries, constitutes a possible source of practical diversity. As the present science communication challenges are not merely European but international in a much wider sense, the appreciation of those differences might even be helpful outside European contexts.

Other parts of the world are often keen to take their cues from Europe or, rather, from 'the West' perceived as a monolith. The message, therefore, that there is no such monolith, that there is diversity, differences and, indeed, disagreement, might leave more opportunities open to other cultures that are struggling to come to terms with the expansion of science and may not be aware that the demand for promises and certainty from science and the expectation that science will actually be able to deliver on such promises, varies not only with time but also – even within Europe – across geographical boundaries. Correspondingly, the understanding of science, represented by scientists, as no less than the epitome of realism constitutes a cultural trait that has traditionally been dominant in some but not all European contexts.[56] The recognition of such differences may prevent the naturalization of cultural traits that function as points of departure for understandings of the mores and means of science communication. Thus, it may serve to preserve a multiplicity of possible approaches connected to diverse understandings of science and its role in society.

All and nobody are to blame for present tendencies to ignore – and thereby waste – the European diversity. Apparently unaware of the implicit normativity of languages, participants in international – and, as a rule, English-spoken – discussions are likely to bring their own specific interpretations of internationally shared notions to the debates and to take for granted that those specific understandings are universally shared. If those understandings, originating partly in different languages, were made explicit, international discussions on universally shared concepts would present unique opportunities for all participants to be enriched with fresh perspectives on their own internalized beliefs and conceptual understandings. Such clarifications seem, however, to be rare. Mostly, the various intuitive interpretations of shared concepts remain implicit and inaccessible to inspection and comparison. The net results are confused debates and the adoption of crude versions of the dominant understandings.

Enlightening tensions and the benefits of contradiction

The presupposition that contradiction is beneficial is the raison d'être behind the classical political institution of public discussion. Disagreement and tensions are taken to be possible sources of knowledge. Processes of opinion formation are seen as processes of enquiry into practical-political problems. The widespread assumption of a science versus politics dichotomy hampers the acknowledgement that closely related presuppositions form part of the foundation of the ethos of science. The internal, scientific communication norm of organized scepticism only makes sense as an appreciation of contradiction. In that case, however, the appreciation is restricted to the fraternity of peers and to arguments of a scientific nature. The consequences of these restrictions are far-reaching. In effect, individual scientists and science as a societal institution are largely deprived of the benefit of contradiction from other points of view and other ways of reasoning.

As a human enterprise, increasingly concerned with practical-political questions, science needs arguments from other positions and perspectives to enable it to recognize and cope with its own limitations. They are not visible from within. Awareness of presuppositions and norms that are present in scientific work can only be achieved by interactions with others who do not share those presuppositions and norms. This is the reason why there is a need to 'ensure the exposure of hypotheses to the broadest range of criticism'.[57] Scientists may not be able to recognize the boundaries of their knowledge if 'the necessary criticism is missing'. There is a real risk, then, that scientific imagination could degenerate into mere fantasies.[58] Such fantasies, in turn, may be communicated to the public at large and, due to the authority of science, be widely recognized as scientific knowledge.

Not all modes of presentation are equally suited to the encouragement of arguments from other positions and perspectives. In dealing with problems of a practical-political nature, it has been suggested, 'the modes of reasoning used in the argument are much richer than those permissible in the scientific discourse'.[59] Such richer – and, thus, less exact – modes of reasoning are, however, incompatible with the forceful and highly persuasive scientific rhetoric of numbers and do not entail a license to make strong knowledge claims. They are richer precisely because they do not pertain to truth-seeking and technical problem solving, but to practical-political problems and, thus, should invite exchanges among different points of view in order to further deliberation.[60] They are richer because they are concerned with questions that can only be resolved by way of discussion.

Richness, thus, comes at the price of decreased persuasive force and authority. Scientific researchers may, however, by truthfully presenting their cases and inviting others to join the arguments, earn a high degree of trust as authoritative voices while achieving some of the benefits of contradiction. Moreover, because presentational style is not merely presentational style[61] but also expresses and may reinforce ways of thinking and identifying, habits of richer presentations of scientific research concerning science-related public affairs and political issues might, in the long-term, facilitate the combination of the identities of scientist and citizen.

In the long-term, the maintenance of societies based on active citizenship is likely to be adversely affected if most citizens adopt the identity of objects-cum-consumers of scientific research, interventions and policies, and scientists take on the role of observers-cum-producers of such interventions and policies. To prevent that from happening it would seem wise to reconsider the ambiguity of the enlightenment motto *sapere aude*. We should dare to know *and* dare to think. Democratic knowledge societies must accommodate both.

The introduction of a political category of science communication can be seen as a possible means to preserve the ambiguity. Suited to practical-political issues and featuring citizens on an equal footing – some of them scientists – who represent different points of view and ways of reasoning and share responsibility for public affairs, it would not be an easy way out of current problems. Based on the classical, Aristotelian distinction between technical-scientific and practical-political issues, it would challenge scientists and other citizens to distinguish, from one case to another, between such issues.

First and foremost, it comes with the assumption that science, as all human activities, has limits and, thus, should not be seen as an all-purpose problem solver. Political problems, rather, should be resolved by political means. There is tension, in other words, between the idea of science as universal light and the proposed category. It does, however, have room for science as a means of answering a multiplicity of questions. It is not out to conquer reality as a whole; there is no crusading mission. And to science as a societal institution and an intellectual enterprise it brings the advantage that it might keep know-ledge societies alive as political entities with that room for a civil society of multiple positions and perspectives that served, in the first place, as a fertile ground for the development of science.

Aims of achieving control cannot be removed from science. They are endemic to the scientific logic. It is a recurrent problem – increasing along with the expansion of science – that all too easily these aims of achieving control get out of hand themselves. They expand to encompass the possible control of humans by humans. There seems to be no other option of response than to keep discussions going on science as a human enterprise, complete with assumptions that can be questioned, with economic and social interests that might need curbing, and with substantial disagreements that it is beyond the ability of science to resolve.

To societies calling themselves knowledge societies and to practitioners of science communication it would seem a particular obligation to stimulate con-tinuous debate on what it is possible to know in what ways and by what means. What significance can and should be attributed, when and why, to specialized knowledge, originating in empirical science? What significance can and should be attributed, when and why, to exchanges among different points of view? And what does it take to maintain science as an intellectual endeavour, open to sceptical questioning and critical thought?

We need a diversity of understandings of science communication, a polit-ical category of discussion among them, to allow the continuous probing of those evergreen key questions. About the limits and the authority of science. And about the relationships between science and political democracy – between truth and disagreement.

Snapshot XVI

Big Data, Algorithms and the Stereotyping of Citizens

In the late 1980s, American citizen Andrew Sokolow committed a crime against statistical profiling. He fulfilled a series of criteria indicating that he was a drug trafficker. Only, he wasn't one. Travelling 10 hours by air to Miami, his destination and cash payment had triggered surveillance activities. He had been observed carrying only hand luggage, staying for just two days in Miami, appearing to be nervous and having given the airline a phone number that did not correspond to the name on his ticket. He was taken to prison. In April 1989, he lost his claim for compensation at the Supreme Court of the United States. A majority of the judges found his arrest warranted by the fact that he had behaved like the stereotype of a drug trafficker.

The Sokolow case – an omen of future statistical disciplining – has been pointed to by Spiros Simitis, a scholar of law who for 16 years served as the world's first data protection commissioner in the German state of Hessen. In the early 1990s, concerned about the future uses of electronic traces, he suggested the protection of anonymity as a guiding principle of the further development of electronic systems. Citizen transparency, he argued, would threaten political democracy.

Since then, big data has become big business for public authorities and private companies. The storage of personal data has become a purpose of electronic transactions in its own right. And an increasing number of professions, scientific specialities and would-be sciences have tucked in to the enterprise of gathering and utilizing such data to create formula – algorithms, based on the identification of patterns in large-scale data sets – for multiple purposes: to carry out financial transactions; to disseminate 'personalized' advertisements, news and medicine; to nudge subjects to behave in certain ways; to carry out preemptive strikes against crimes and diseases, creating groups of pre-patients and pre-criminals in the process; to predict educational dropout; to decide on requests for prison paroles, and so on.

The hybrid science and profession of Information Technology (IT) is there, of course – and, thus, mathematics and physics – but so are economists, epidemiologists, geneticists – and, thus, biology – medics in general, psychologists, sociologists, statisticians, et cetera. A large number of sciences seem to be fusing into a societal machinery of behavioural prediction and control, founded on the steady collection of trillions of electronic traces, feeding into probability calculations and statistical profiling. Possibly, benefits (to some) of that kind of development will find their own way. To deal with side effects, questionable benefits and the possible revival of old orthodoxies and mythologies in new guises, however, there is a need for scientists to engage in critical discussions across disciplinary borders. There is also a chance that such discussions may actually unfold.

The development has been accompanied by science understandings and ambitions that go at least two centuries back in time – back to the times of 'moral statistics' and 'social physics' – and quite a few scientists may currently be experiencing a blow to their own understandings and aspirations. The return of behaviourism with its machine-like view of humans. The reappearance of beliefs in the unlimited predictive power of science, seen as a non-interpretative fact-producing activity. The increasing closeness to very big money with its affinity for secrecy, not least about algorithms and their background. This prepares the ground for a multitude of careers, but it does not represent everybody's science idea(l)s. Some will take a critical stance.

It is a possible positive side effect of the development in China of a digital surveillance regime – complete with the creation of categories of subjects to be punished or rewarded according to their social scores – that it may spur critique in other parts of the world. The myriad of thorny questions to be faced may serve – another positive side effect – to call forth thoughtful rather than merely PR-minded scientists. Uncertainties, basic assumptions and values and the influence of vested interests belong on the agenda. What understandings of the good life lie beneath various surveillance schemes? What assumptions about humans lie beneath various algorithms? How may current developments of artificial intelligence mould understandings of human intelligence, confusing 'computation' and 'thought' in the process? On a more specific note: How is knowledge of groups transferred to the level of individuals? Is there a risk of an automation of social prejudices and of the systematic stigmatization of certain social groups? How, then, did those groups come into existence – what kind of reality should be ascribed to them outside the social-scientific imagination? And are we, by the way, really able to get out of social reality, to observe and operate it from the outside?

In the 1960s, Donald M. MacKay (1922–1987), physicist and early information theorist, shrewdly observed that the publication of behavioural predictions was bound to make them self-fulfilling: 'If any future use of computers wants watching on behalf of mankind, it is this; for our society's insatiable thirst for information about itself and its future has now laid it wide open to the most subtle bondage of all, in which major decisions can in principle be taken for it (wittingly or otherwise) by those whom it asks to predict them', MacKay argued, warning against the possible future manipulation of attitudes 'under the guise of scientific prediction'.

More recently, the predictive-cum-prescriptive uses of big data have been ascribed the potential to substitute compliance and adaptability for the capacity for independent and critical thought. We had better be quick before too many turn to obeying the stereotypes. Political discussions about digitization and the collection and use of personal data depend on the participation of thoughtful scientists and other citizens.

The textual snapshot about big data, algorithms and the stereotyping of citizens has drawn on Gitte Meyer, 'Kreditkortet og demokratiet'; Spiros Simitis, 'Privacy: An Endless Debate'; Armand Mattelart, *The Invention of Communication*, 227–29; Donald M. MacKay, 'Machines and Societies', 164; and Evgeny Morozov, 'The Real Privacy Problem'. See also Stefan Strauss, 'If I Only Knew Now What I Know Then. . .'; Robindra Prabhu, 'How Should We Govern the Algorithms That Shape Our Lives?'; and danah boyd and Kate Crawford, 'Critical Questions for Big Data'.

Notes

1 Aristotle, *The Nicomachean Ethics*, 1127a20.
2 Ibid., 1127b15.
3 Helga Nowotny, Peter Scott and Michael Gibbons, *Re-Thinking Science: Knowledge and the Public in an Age of Uncertainty*, 38.
4 For a discussion of the concept of hype, see Gitte Meyer, 'Expectations and Beliefs in Science Communication: Learning from Three European Gene Therapy Discussions of the Early 1990s'.
5 Hans Primas, explores 'Fascination and Inflation in Science'.
6 Adam Bly, ed., *Science Is Culture*, ix.
7 The remarks about widespread modes of research presentation are based on conference observations. For a more thorough characterization, see Gitte Meyer, *Lykkens kontrollanter: Trivselsmålinger og lykkeproduktion* [The happiness controllers: The measurement of well-being and the production of happiness].
8 Gitte Meyer and Peter Sandøe, 'Going Public: Good Scientific Conduct'.
9 Plots of televised crime fiction are quite often constructed around the theme of corruption in science – medical science, in particular – but the topic seems to be shunned in the day-to-day public exchanges on science in society unless it proves useful as part of the armour in a science war. Thus, nuanced discussions of how financial interests in science might affect the ways scientific enquiries are carried out, presented and put to use are in short supply.
10 Among numerous possible examples, see for instance Ulrich Beck, *Risk Society: Towards a New Modernity*; Sharon M. Friedman, Sharon Dunwoody and Carol L. Rogers, eds., *Communicating Uncertainty: Media Coverage of New and Controversial Science*.
11 Quoted in Meyer and Sandøe, 'Going Public'.
12 A rather early example of the interpretation of disagreement among scientists as a symptom of scientific uncertainty can be found in Sharon Dunwoody, 'Scientists, Journalists, and the Meaning of Uncertainty'.
13 S. Holly Stocking, 'How Journalists Deal with Scientific Uncertainty'.
14 BBC Trust, 'BBC Trust Review of Impartiality and Accuracy of the BBC's Coverage of Science.
15 BBC Trust, 'BBC Trust Review of Impartiality and Accuracy of the BBC's Coverage of Science: Follow Up', 1.
16 Ibid.
17 Ibid., 8.
18 Robert K. Barnhart, ed., *Dictionary of Etymology*.
19 Quoted in Jürgen Habermas, *Borgerlig offentlighet: dens framvekst og forfall: henimot en teori om det borgerlige samfunn* [The structural transformation of the public sphere: An inquiry into a category of bourgeois society], 88.
20 Ibid., 87.
21 Aristotle, *The Politics*, 1261a10, 1263a40; Crick, *In Defence of Politics*.
22 Hannah Arendt, *The Human Condition*, 7.
23 Walter Lippmann, *Public Opinion*; John Stuart Mill, 'On Liberty'.
24 Hans Magnus Enzensberger, *Einzelheiten I: Bewusstseins-Industrie*.
25 Bob Ward, 'Reporters Feel the Heat over Climate Change'.
26 Maxwell T. Boykoff and S. Ravi Rajan, 'Signals and Noise: Mass-Media Coverage of Climate Change in the US and the UK'.

27   Chris Mooney, 'Blinded by Science: How "Balanced" Coverage Lets the Scientific Fringe Hijack Reality'.

28   Julia B. Corbett and Jessica L. Durfee, 'Testing Public (Un)certainty of Science: Media Representations of Global Warming'. For convenience, all examples are picked from the debate about human-made climate change, but a whole range of other science-related public affairs might have been equally relevant.

29   Robert K. Merton, *Social Theory and Social Structure*, 613–14.

30   Ibid., 596.

31   Harold Stearns, *America and the Young Intellectual*, 105–6.

32   Hans-Georg Gadamer, *Truth and Method*, 552.

33   Quoted in Gitte Meyer, *Offentlig fornuft? Videnskab, journalistik og samfundsmæssig praksis* [Public reason? Knowledge, journalism and societal practice].

34   Austin L. Hughes, 'The Folly of Scientism'.

35   Barnhart, *Dictionary of Etymology*.

36   *Duden: Das Herkunftswörterbuch*.

37   *Ordbog over det danske sprog* [Dictionary of the Danish language].

38   Barnhart, *Dictionary of Etymology*. My italics, gm.

39   Gadamer, *Truth and Method*, 552.

40   The unwillingness to recognize limitations and the tendency to think instead in terms of 'a consistency that exists nowhere in reality' has been tied to totalitarian frameworks of thought. See Hannah Arendt, *The Origins of Totalitarianism*, 607.

41   Adam Ferguson, *An Essay on the History of Civil Society*, pt. 4 sec. 1.

42   Ronald Barnett, *Realizing the University in an Age of Supercomplexity*, 83, argued that '[t]he overall challenge amounts to keeping the ladder of the internal values in place, not kicked away'.

43   John Ziman, *Real Science*, 78–79, 81.

44   Merton, *Social Theory and Social Structure*, 614.

45   Stephen Schneider, 'Don't Bet All Environmental Changes Will Be Beneficial'.

46   I owe the distinction between research *on* human objects and *with* human agents to Mary Midgley, *Science as Salvation: A Popular Myth and Its Meaning*, 209.

47   Aristoteles, *Retorik* [Rhetoric].

48   Gadamer, *Truth and Method*, 568.

49   J. D. G. Evans, *Aristotle's Concept of Dialectic*.

50   Some components of a dialectical approach to science communication were unfolded – not under that heading, though – by the head of the science division of Sveriges Radio (Swedish public service broadcaster), Ulrika Björkstén in 2012, 'Expertsamhället riskerar att bidra till fördumning' [The expert society may lead to a state of stupidity]. As her point of departure, Björkstén used concerns about democracy in the light of the rise of the 'expert society', combining increasing specialization in science and an increased significance of science in and to society. The maintenance of a democratic society, she found, depends on the maintenance of critical, competent and independent science journalism, neither assuming science to be good nor bad by definition, but enabling scientific knowledge claims to be presented, enquired into and discussed openly in public. Otherwise, she feared, society might be pushed towards 'a state of stupidity'.

51   For a more than 50-year-old, but not necessarily outdated discussion of the responsibility of scientists, see Max Born, *Von der Verantwortung des Naturwissenschaftlers*.

52   Dorothy Nelkin and M. Susan Lindee, *The DNA Mystique: The Gene as a Cultural Icon*, 128.

53 Sheila Jasanoff, *Designs on Nature: Science and Democracy in Europe and the United States.*
54 Ibid., 255.
55 Ibid.
56 See Meyer, 'Expectations and Beliefs in Science Communication' for a cross-cultural comparison of basic attitudes to and expectations of science and scientists, expressed in British, Danish and German newspapers during the gene therapy debates of the early 1990s.
57 Helen E. Longino, *The Fate of Knowledge*, 165, 132.
58 Primas, 'Fascination and Inflation in Science', 86, 79.
59 Horst Rittel and Melvin Webber, 'Dilemmas in a General Theory of Planning', called problems of a practical-political nature 'wicked problems'.
60 Sarah Atkinson and Kerry E. Joyce, 'The Place and Practices of Well-Being in Local Governance', made this point with respect to the concept of well-being.
61 The relationships between presentational styles and identity are discussed by Hans Magnus Enzensberger, *Fortuna und Kalkül: Zwei mathematische Belustigungen*, 39–40; and Martha C. Nussbaum, *The Fragility of Goodness: Luck and Ethics in Greek Tragedy and Philosophy*, 122–35.

BIBLIOGRAPHY

Altschull, J. Herbert. *From Milton to McLuhan: The Ideas Behind American Journalism*. New York and London: Longman, 1990.

Arasse, Daniel. *Bildnisse des Teufels*. Berlin: Matthes & Seitz, 2012.

Arendt, Hannah. *The Human Condition*. Chicago and London: University of Chicago Press, 1969 [1957].

———. 'Kultur und Politik'. *Merkur, Deutsche Zeitschrift für Europäisches Denken* 12 (1958–59): 1122–45.

———. *The Origins of Totalitarianism*. New York: Schocken Books, 2004 [1951].

———. Sonning Prize acceptance speech 1975. Copenhagen, 18 April 1975. Accessed 23 March 2017. http://miscellaneousmaterial.blogspot.com/2011/08/hannah-arendt-sonning-prize-acceptance.html.

Aristoteles. *Retorik* [Rhetoric]. Translated into Danish by Thure Hastrup. København: Museum Tusculanums Forlag, 2002.

Aristotle. *The Nicomachean Ethics*. Translated by J. A. K. Thomson. Revised with notes and appendices by Hugh Tredennick. London: Penguin Books, 2004.

———. *The Politics*. Translated by T. A. Sinclair, revised and re-presented by Trevor J. Saunders. London: Penguin Books, 1992.

Ashley, Maurice. *England in the Seventeenth Century*. London: Penguin Books, 1967.

Assmann, Jan. *The Price of Monotheism*. Stanford, CA: Stanford University Press, 2010.

Atkinson, Sarah, and Kerry E. Joyce. 'The Place and Practices of Well-being in Local Governance'. *Environment and Planning C: Government and Policy* 29 (2011): 133–48. http://doi.org/10.1068/c09200.

Averbeck, Stefanie, and Arnulf Kutsch. 'Thesen zur Geschichte der Zeitungs- und Publizistikwissenschaft 1900–1960'. In *Die Spirale des Schweigens: Zum Umgang mit der nationalsozialistischen Zeitungswissenschaft*, edited by Wolfgang Duchkowitsch, Fritz Hausjell and Bernd Semred, 55–66. Berlin, Hamburg and Münster: LIT Verlag, 2004.

Bahr, Ehrhard, ed. *Was ist Aufklärung? Thesen und Definitionen*. Stuttgart: Philip Reclam jun., 2002 [1974].

Barnett, Ronald. *Realizing the University in an Age of Supercomplexity*. Buckingham and Philadelphia, PA: The Society for Research into Higher Education and Open University Press, 2000.

Barnhart, Robert K., ed. *Dictionary of Etymology*. Edinburgh and New York: Chambers, 2006.

Bauer, Martin W. 'The Evolution of Public Understanding of Science: Discourse and Comparative Evidence'. *Science, Technology and Society* 14 (2009): 221–40. https://doi.org/10.1177/097172180901400202.

BBC Trust. 'BBC Trust Review of Impartiality and Accuracy of the BBC's Coverage of Science'. July 2011. Accessed 24 March 2017. http://downloads.bbc.co.uk/bbctrust/assets/files/pdf/our_work/science_impartiality/science_impartiality.pdf.

————. 'BBC Trust Review of Impartiality and Accuracy of the BBC's Coverage of Science: Follow Up'. November 2012. Accessed 24 March 2017. http://downloads. bbc.co.uk/bbctrust/assets/files/pdf/our_work/science_impartiality/science_impartiality_followup.pdf.

Beard, George M. *American Nervousness: Its Causes and Consequences; A Supplement to Nervous Exhaustion (Neurasthenia)*. New York: G. P. Putnam's Sons, 1881.

Beck, Ulrich. *Risk Society: Towards a New Modernity*. New Delhi: Sage, 1992 [1986].

Bernal, John Desmond. *The Social Function of Science*. London: George Routledge & Sons, 1946 [1939].

Björkstén, Ulrika. 'Expertsamhället riskerar att bidra till fördumning' [The expert society may lead to a state of stupidity]. *Svenska Dagbladet*, 16 October 2012. Accessed 24 March 2017. http://www.svd.se/kultur/understrecket/expertsamhallet-riskerar-att-bidra-till-fordumning_7584858.svd.

Bloom, Allan. *The Closing of the American Mind*. New York, London, Toronto, Sydney and Tokyo: Simon & Schuster, 1987.

Bly, Adam, ed. *Science Is Culture*. New York, London, Toronto, Sydney, New Delhi and Auckland: Harper Perennial, 2010.

Bodmer, Walter. 'Public Understanding of Science: The BA, the Royal Society and COPUS'. *Notes and Records of the Royal Society* 64 (2010): S151–61. http://doi.org/10.1098/rsnr.2010.0035.

Bohrmann, Hans. 'Als der Krieg zu Ende war: Von der Zeitungswissenschaft zur Publizistik'. In *Die Spirale des Schweigens: Zum Umgang mit der nationalsozialistischen Zeitungswissenschaft*, edited by Wolfgang Duchkowitsch, Fritz Hausjell and Bernd Semred, 97–122. Berlin, Hamburg and Münster: LIT Verlag, 2004.

Bon, Gustave le. *The Crowd: A Study of the Popular Mind*. New York: Macmillan, 1896. Accessed 2 May 2017. https://archive.org/details/crowdastudypopu00bongoog.

Born, Max. *Von der Verantwortung des Naturwissenschaftlers*. München: Nymphenburger Verlagshandlung, 1965.

Bottomore, Tom B. *Elites and Society*. Harmondsworth, Baltimore MD and Ringwood: Pelican Books, 1971.

boyd, danah, and Kate Crawford. 'Critical Questions for Big Data: Provocations for a Cultural, Technological, and Scholarly Phenomenon'. *Information, Communication & Society*, 15 (2012): 662–79. http://doi.org/10.1080/1369118X.2012.678878.

Boykoff, Maxwell T., and S. Ravi Rajan. 'Signals and Noise: Mass-Media Coverage of Climate Change in the US and the UK'. *EMBO Reports* 8 (2007): 207–11. http://doi.org/10.1038/sj.embor.7400924.

Bradley, Ryan. 'Why NASA Helped Ridley Scott Create "The Martian" Film'. *Popular Science*, 19 August 2015. Accessed 5 April 2017. http://www.popsci.com/why-nasa-helped-ridley-scott-create-martian-film-and-what-means-future-sci-fi-space-movies.

Burke, Peter. *A Social History of Knowledge: From Gutenberg to Diderot*. Cambridge and Malden, MA: Polity, 2008.

Bury, John B. *The Idea of Progress: An Inquiry into Its Origin and Growth*. New York: Dover Publications, 1955 [1932].

Butterfield, Herbert. *The Origins of Modern Science: 1300–1800*. New York: Macmillan, 1959 [1949]. Accessed 24 March 2017. https://ia802300.us.archive.org/27/items/originsofmoderns007291mbp/originsofmoderns007291mbp.pdf.

Callon, Michel. 'Some Elements of a Sociology of Translation: Domestication of the Scallops and the Fishermen in St Brieuc Bay'. In *Power, Action and Belief: A New Sociology of Knowledge?*, edited by John Law, 196–223. London: Routledge, 1986.

Carey, John. *The Intellectuals and the Masses: Pride and Prejudice among the Literary Intelligensia 1880–1939*. London: Faber and Faber, 1992.

Cater, Douglass. *The Fourth Branch of Government*. New York: Vintage Books, 1959.

Chargaff, Erwin. How Scientific Papers Are Written. *Fachsprache/Special Language* 8 (1986): 106–10.

Cohen, Jean L., and Andrew Arato. *Civil Society and Political Theory*. Cambridge, MA, and London: MIT Press, 1992.

College of Physicians of Philadelphia, *The History of Vaccines*, updated 2017. Accessed 5 April 2017. http://www.historyofvaccines.org/.

Collier, David, Fernando Daniel Hidalgo and Andra Olivia Maciuceanu. 'Essentially Contested Concepts: Debates and Applications'. *Journal of Political Ideologies* 11 (2006): 211–46. http://doi.org/10.1080/13569310600923782.

Coltman, Irene. *Private Men and Public Causes: Philosophy and Politics in the English Civil War*. London: Faber and Faber, 1962.

Converse, Philip E. 'The Nature of Belief Systems in Mass Publics (1964)'. *Critical Review* 18, no. 1–3 (2006): 1–74. http://doi.org/10.1080/08913810608443650.

Cooper, Anthony Ashley, Third Earl of Shaftesbury. 'A Letter Concerning Enthusiasm'. In *Characteristicks of Men, Manners, Opinions, Times*, Vol. 1, 1737, edited by Douglas den Uyl. Indianapolis: Liberty Fund, 2001. Accessed 24 March 2017. http://oll.libertyfund. org/titles/shaftesbury-characteristicks-of-men-manners-opinions-times-vol-1#lf5987_ head_005.

Cooter, Roger, and Stephen Pumfrey. 'Separate Spheres and Public Places: Reflections on the History of Science Popularization and Science in Popular Culture'. *History of Science* 32 (1994): 237–67. https://doi.org/10.1177/007327539403200301.

Corbett, Julia B., and Jessica L. Durfee. 'Testing Public (Un)certainty of Science: Media Representations of Global Warming'. *Science Communication* 26 (2004): 129–51. http://doi.org/10.1177/1075547004270234.

Crick, Bernard. *The American Science of Politics: Its Origins and Conditions*. Berkeley and Los Angeles: University of California Press, 1964 [1959].

———. *In Defence of Politics*. London and New York: Continuum, 2005 [1962].

Dahl, Robert A. *A Preface to Democratic Theory*. Chicago and London: University of Chicago Press, 1966 [1956].

Demos, British Council, SNS and The UK Presidency of the EU 2005. *The Network Effect: Connecting Europe's Next Generation Leaders: Media and Legitimacy in European Democracy*. Stockholm, 2005.

Dewey, John. *The Public and Its Problems*. Athens: Swallow Press and Ohio University Press, 1991 [1927].

Duden: Das Bedeutungswörterbuch. 3. Auflage. Mannheim, Leipzig, Wien and Zürich: Dudenverlag, 2002.

Duden: Das Herkunftswörterbuch. 4. Auflage. Mannheim, Leipzig, Wien and Zürich: Dudenverlag, 2007.

Dunwoody, Sharon. 'Scientists, Journalists, and the Meaning of Uncertainty'. In *Communicating Uncertainty: Media Coverage of New and Controversial Science*, edited by Sharon M. Friedman, Sharon Dunwoody and Carol L. Rogers, 59–79. Mahwah, NJ and London: Lawrence Erlbaum, 1999.

Eijk, Dick van, ed. *Investigative Journalism in Europe*. Amsterdam: Vereniging van Ondersoeksjournalisten VVOJ, 2005.

Einsiedel, Edna. 'Editorial: Of Publics and Science'. *Public Understanding of Science* 16 (2007): 5–6. http://doi.org/10.1177/0963662506071289.

Enzensberger, Hans Magnus. *Einzelheiten I: Bewusstseins-Industrie.* Frankfurt am Main: Suhrkamp Verlag, 1964.

———. *Fortuna und Kalkül: Zwei matematische Belustigungen.* Frankfurt am Main: Suhrkamp Verlag, 2009.

European Commission. *Integrating Science in Society Issues in Scientific Research: Main Findings of the Study on the Integration of Science and Society Issues in the Sixth Framework Programme.* EUR 22976. Brussels: European Commission, 2007.

———. *Mid-Term Assessment: Science and Society Activities 2002–2006: Final Report 22 March 2007.* EUR 22954. Brussels: European Commission, 2007.

———. *Public Engagement in Science.* EUR 23334. Luxembourg: Office for Official Publications of the European Communities, 2008. http://doi.org/10.2777/20800.

Evans, J. D. G. *Aristotle's Concept of Dialectic.* Cambridge, London, New York and Melbourne: Cambridge University Press, 2010 [1977].

Everding, Gerry. 'Genetically Modified Golden Rice Falls Short on Lifesaving Promises'. *The Source,* 2 June 2016. Washington University in St. Louis. Accessed 5 April 2017. https://source.wustl.edu/2016/06/genetically-modified-golden-rice-falls-short-lifesaving-promises/.

Felt, Ulrike, ed. *O.P.U.S. Optimising Public Understanding of Science and Technology: Final Report.* Vienna: Department for Philosophy of Science and Social Studies of Science, University of Vienna, 2003.

Ferguson, Adam. *An Essay on the History of Civil Society.* 1767. Accessed 2 May 2017. http://socserv2.socsci.mcmaster.ca/~econ/ugcm/3ll3/ferguson/civil1.

Fink, Leon. *Progressive Intellectuals and the Dilemmas of Democratic Commitment.* Cambridge, MA, and London: Harvard University Press, 1997.

Friedman, Sharon M., Sharon Dunwoody and Carol L. Rogers, eds. *Communicating Uncertainty: Media Coverage of New and Controversial Science.* Mahwah, NJ, and London: Lawrence Erlbaum, 1999.

———. *Scientists and Journalists: Reporting Science as News.* New York and London: Free Press, 1986.

Fukuyama, Francis. *The End of History and the Last Man.* London: Penguin Books, 1992.

Furedi, Frank. *Where Have All the Intellectuals Gone? Confronting 21st Century Philistinism.* London and New York: Continuum, 2004.

Gadamer, Hans-Georg. *Truth and Method,* 2nd, rev. ed. London: Sheed & Ward, 2001 [1975].

George, Dorothy. *England in Transition: Life and Work in the Eighteenth Century.* London: Penguin Books, 1953.

Gerhardt, Volker. *Partizipation: Das Prinzip der Politik.* München: C. H. Beck, 2007.

Giddens, Anthony. *Beyond Left and Right: The Future of Radical Politics.* Cambridge: Polity Press, 1994.

Gieryn, Thomas F. *Cultural Boundaries of Science: Credibility on the Line.* Chicago and London: University of Chicago Press, 1999.

Gilcher-Holtey, Ingrid. *Die 68er Bewegung: Deutschland, Westeuropa, USA.* München: C. H. Beck, 2005.

Glasser, Theodore L., ed. *The Idea of Public Journalism.* New York and London: Guilford, 1999.

Goede, Wolfgang C. *Civil Journalism & Scientific Citizenship: Scientific Communication 'of the People, by the People and for the People'.* Keynote address to the Third World Conference of Science Journalists. 2002. Accessed 31 March 2017. http://comm-org.wisc.edu/papers2003/degoede.htm.

Habermas, Jürgen. *Borgerlig offentlighet: dens framvekst og forfall: henimot en teori om det borgerlige samfunn* [The structural transformation of the public sphere: An inquiry into a category of bourgeois society]. Oslo: Fremad/Gyldendal Norsk Forlag, 1980 [1962].

———. 'Technik und Wissenschaft als "Ideologie"?' *Man and World* 1 (1968): 483–523.

Haldane, J. B. S. 'Biological Possibilities for the Human Species in the Next Ten Thousand Years'. In *Man and His Future*, edited by Gordon Wolstenholme, 337–61. London: J. & A. Churchill, 1963.

Hallin, Daniel C., and Paolo Mancini. *Comparing Media Systems: Three Models of Media and Politics*. Cambridge: Cambridge University Press, 2004.

Hardt, Hanno. 'Am Vergessen scheitern: Essay zur historischen Identität der Publizistikwissenschaft, 1945–68'. In *Die Spirale des Schweigens: Zum Umgang mit der nationalsozialistischen Zeitungswissenschaft*, edited by Wolfgang Duchkowitsch, Fritz Hausjell and Bernd Semred, 153–60. Berlin, Hamburg and Münster: LIT Verlag, 2004.

Hayden, Tom, and Dick Flacks. 'The Port Huron Statement at 40'. *The Nation*, 5 August 2002. Accessed 31 March 2017. http://www.thenation.com/article/port-huron-statement-40.

Hentschel, Manfred W. 'Wir fordern die Enteignung Axel Springers'. Spiegel-Gespräch mit dem FU-Studenten Rudi Dutschke (SDS). *Der Spiegel* 29 (1967): 29–33.

Hilgartner, Stephen. 'The Dominant View of Popularization: Conceptual Problems, Political Uses'. *Social Studies of Science* 20 (1990): 519–39. https://doi.org/10.1177/030631290020003006.

Hill, Christopher. *The Century of Revolution: 1603–1714*. London and New York: Routledge, 2010 [1961].

Hobsbawm, Eric. *Age of Extremes: The Short Twentieth Century 1914–1991*. London: Abacus, 1995.

Höffe, Otfried. *Thomas Hobbes*. München: Verlag C. H. Beck, 2010.

Hofstadter, Richard. *Anti-intellectualism in American Life*. New York: Vintage Books, 1962–63.

———. 'The Paranoid Style in American Politics'. *Harper's Magazine* (November 1964): 77–86.

Hornby, A. S., ed. *Oxford Advanced Learner's Dictionary of Current English*, 5th ed. Oxford: Oxford University Press, 1995.

———. *Oxford Advanced Learner's Dictionary of Current English*, 8th ed. Oxford: Oxford University Press, 2010.

Horst, Maja: 'Public Expectations of Gene Therapy: Scientific Futures and Their Performative Effects on Scientific Citizenship'. *Science, Technology & Human Values* 32 (2007): 150–71. https://doi.org/10.1177/0162243906296852.

Hourani, Albert. *A History of the Arab Peoples*. London: Faber and Faber, 2002 [1991].

Hughes, Austin L. 'The Folly of Scientism'. *The New Atlantis* 37 (Fall 2012): 32–50.

Huxley, Julian. *Memories*. Harmondsworth and Ringwood: Penguin Books, 1972.

———. *Memories II*. Harmondsworth: Penguin Books, 1978.

Jacob, Margaret C. *The Enlightenment: A Brief History with Documents*. Boston and New York: Bedford/St.Martin's, 2001.

———. *The Radical Enlightenment: Pantheists, Freemasons and Republicans*. Lafayette, LA: Cornerstone, 2006.

Jacoby, Russell. *The Last Intellectuals: American Culture in the Age of Academe*. New York: Basic Books, 2000 [1987].

Jasanoff, Sheila. *Designs on Nature: Science and Democracy in Europe and the United States*. Princeton, NJ, and Oxford: Princeton University Press, 2005.

Kahlor, LeeAnn, and Patricia A. Stout, eds. *Communicating Science: New Agendas in Communication*. New York and London: Routledge, 2010.

Kapuściński, Ryszard. *Notizen eines Weltbürgers*. Berlin: Eichborn, 2002.

Kleinsteuber, Hans J. 'Habermas and the Public Sphere: From a German to a European Perspective'. *Javnost: The Public* 8 (2001): 95–108. http://doi.org/10.1080/13183222.2001.11008767.

Koselleck, Reinhart. *Begriffsgeschichten*. Frankfurt am Main: Suhrkamp Verlag, 2006.

Kracauer, Siegfried. *Die Angestellten*. Frankfurt am Main: Suhrkamp, 1971 [1929].

Kristeller, Paul Oskar. *Renaissance Thought: The Classic, Scholastic and Humanist Strains*. New York, Evanston, IL and London: Harper Torchbooks, 1961.

Lang, Markus. 'Der Marktplatz: Ort der entpolitisierten Öffentlichkeit'. In *Weimar als politische Kulturstadt*, edited by Klaus Dicke and Michael Dreyer, 65–72. Berlin: Verlag Jena 1800, 2006.

Latour, Bruno. 'On Interobjectivity'. *Mind, Culture, and Activity* 3 (1996): 228–45. http://doi.org/10.1207/s15327884mca0304_2.

Lederberg, Joshua. 'Biological Future of Man'. In *Man and His Future*, edited by Gordon Wolstenholme, 263–73. London: J. & A. Churchill, 1963.

Lévy-Leblond, Jean-Marc. 'About Misunderstandings about Misunderstandings'. *Public Understanding of Science* 1 (1992): 17–21. https://doi.org/10.1088/0963-6625/1/1/004.

Liessmann, Konrad Paul. *Lob der Grenze: Kritik der politischen Unterscheidungskraft*. Wien: Paul Zsolnay Verlag, 2012.

Lippmann, Walter. *Public Opinion*. New York: Simon & Schuster, 1997 [1922].

Lipset, Seymour Martin. *Political Man: The Social Bases of Politics*. Garden City, NY: Doubleday, 1960.

Longino, Helen E. *The Fate of Knowledge*. Princeton, NJ, and Oxford: Princeton University Press, 2002.

Ludwig, Jeff. 'From Apprentice to Master: Christopher Lasch, Richard Hofstadter, and the Making of History as Social Criticism'. *Essays in History*, 2011. Accessed 31 March 2017. http://www.essaysinhistory.com/articles/2011/30.

MacIntyre, Alasdair. *After Virtue: A Study in Moral Theory*. Notre Dame, IN: University of Notre Dame Press, 1984.

MacKay, Donald M. 'Machines and Societies'. In *Man and His Future*, edited by Gordon Wolstenholme, 153–167. London: J. & A. Churchill, 1963.

Marcuse, Herbert. *One-Dimensional Man*. Boston: Beacon, 1968.

Marquard, Odo. *Skepsis in der Moderne: Philosophische Studien*. Stuttgart: Philipp Reclam jun., 2007.

Mattelart, Armand. *The Invention of Communication*. Minneapolis and London: University of Minnesota Press, 1996.

McDougall, William. *The Group Mind*, 2nd ed. London: Cambridge University Press, 1927.

McNeil, Maureen. 'Between a Rock and a Hard Place: The Deficit Model, the Diffusion Model and Publics in STS'. *Science as Culture* 22 (2013): 589–608. http://doi.org/10.1080/14636778.2013.764068.

Merton, Robert K. *Social Theory and Social Structure*, enlarged ed. New York and London: Free Press & Collier-Macmillan, 1968 [1949].

Mettrie, Julien Offray de la. 'Man a Machine' [1747]. Excerpt in *The Portable Enlightenment Reader*, edited by Isaac Kramnick, 202–9. New York: Penguin Books, 1995.

Meyer, Gitte. *Den kunstige krop* [The artifical body]. København: Munksgaard, 1991.

——. 'Encounters between Science Communication Idea(l)s: A Comparative Exploration of Two Science Communication Logics, with a Focus on Possible Conflicts and Potential for Mutual Learning'. In *Ethical Issues in Science Communication: A Theory-Based Approach*, edited by Jean Goodwin, Michael F. Dahlstrom and Susana Priest, 173–85. Charleston, SC: CreateSpace, 2013.

——. 'Expectations and Beliefs in Science Communication: Learning from Three European Gene Therapy Discussions of the Early 1990s'. *Public Understanding of Science* 25 (2016): 317–31. http://doi.org/10.1177/0963662514552955.

——. 'Fascinating! Popular Science Communication and Literary Science Fiction: The Shared Features of Awe and Fascination and Their Significance to Ideas of Science Fictions as Vehicles for Critical Debate about Scientific Enterprises and Their Ethical Implications'. In *Science Fiction, Ethics and the Human Condition*, 59–80, edited by Christian Baron, Nicolai Halvorsen and Christine Cornea. Springer International, 2017.

——. 'Gylden ris har lang rejse foran sig' [Golden rice has a long journey ahead]. *Genetik i praksis* 2 no. 3 (2001): 1–3. Danish Centre for Bioethics and Risk Assessment.

——. *Hjernen og eftertanken* [Brains and reflections]. HjerneÅret [The Year of the Brain], 1997.

——. 'Knald eller fald for organer fra grise' [Neck or nothing for organs from pigs]. *Fra rådet til tinget* [From the board to the parliament] no. 117, June 1998. Danish Board of Technology.

——. 'Kreditkortet og demokratiet' [Credit cards and democracy]. Interview with Spiros Simitis in *Djøfbladet* [journal published by the Danish Association of Lawyers and Economists], Summer 1995.

——. *Lykkens kontrollanter: Trivselsmålinger og lykkeproduktion – og videnskab og politik* [The happiness controllers: The measurement of well-being and the production of happiness – and science and politics]. København: Djøf-forlag, 2016.

——. *Offentlig fornuft? Videnskab, journalistik og samfundsmæssig praksis* [Public reason? Knowledge, journalism and societal practice], PhD dissertation. Odense: Syddansk Universitetsforlag, 2005.

——. 'In Science Communication, Why Does the Idea of a Public Deficit Always Return?' *Public Understanding of Science* 25 (2016): 433–46. http://doi.org/10.1177/0963662516629747.

——. 'Scientists, Other Citizens, and the Art of Practical Reasoning'. In *Between Scientists & Citizens: Proceedings of a Conference at Iowa State University, June 1–2, 2012*, edited by Jean Goodwin, 297–306. Ames, IA: Great Plains Society for the Study of Argumentation, 2012.

——. *Why Clone Farm Animals? Goals, Motives, Assumptions, Values and Concerns among European Scientists Working with Cloning of Farm Animals*. Danish Centre for Bioethics and Risk Assessment. Project report 8, 2005. Accessed 5 April 2017. http://curis.ku.dk/ws/files/91331855/WHY_CLONE_FARM.pdf.

Meyer, Gitte, and Anker Brink Lund. 'Almost Lost in Translation: Tale of an Untold Tradition of Journalism'. In *Retelling Journalism: Conveying Stories in a Digital Age*, edited by Marcel Broersma and Chris Peters, 27–46. Leuven: Peeters, 2014.

——. 'International Language Monism and Homogenisation of Journalism'. *Javnost: The Public* 15, no. 4 (2008): 73–86. http://doi.org/10.1080/13183222.2008.11008983.

——. 'Klimadiskussionens diskussionsklima: Polarisering i den offentlige debat om klimaændringer' [The debating climate of the climate debate: Polarization in the public debate about climate change]. In *Jorden brænder: Klimaforandringerne i videnskabsteoretisk og etisk perspektiv* [Earth on fire: Climate change from a philosophical and ethical perspective], edited by Mickey Gjerris, Christian Gamborg, Jørgen E. Olesen and Jakob Wolf, 127–50. Frederiksberg: Forlaget Alfa, 2009.

Meyer, Gitte, and Peter Sandøe. 'Going Public: Good Scientific Conduct'. *Journal of Science and Engineering Ethics* 18 (2012): 173–97. http://doi.org/10.1007/s11948-010-9247-x.

Midgley, Mary. *Science as Salvation: A Popular Myth and Its Meaning*. London and New York: Routledge, 1992.

Mill, John Stuart. 'On Liberty'. In *John Stuart Mill: On Liberty and Other Essays*, edited by John Gray, 1–128. Oxford: Oxford University Press, 1998 [1859].

Ministry of Science, Innovation and Higher Education. *Science in Dialogue. Towards a European Model for Responsible Research and Innovation*. Odense, Denmark, 23–25 April 2012: Conference programme.

Montaigne, Michel de. *Essais*, erste moderne Gesamtübersetzung von Hans Stilett. Frankfurt am Main: btb, 2000.

Mooney, Chris. 'Blinded by Science: How "Balanced" Coverage Lets the Scientific Fringe Hijack Reality'. *Columbia Journalism Review* 43, no. 4 (2004). Re-published in *Discover*, 15 January 2010. Accessed 5 April 2017. http://blogs.discovermagazine.com/intersection/2010/01/15/blinded-by-science-how-balanced-coverage-lets-the-scientific-fringe-hijack-reality/#.WOS7LG_yjX4.

Morgan, Edmund S. *Inventing the People: The Rise of Popular Sovereignty in England and America*. New York and London: W. W. Norton, 1989.

Morozov, Evgeny. 'The Real Privacy Problem'. *MIT Technology Review*, October 22, 2013. https://www.technologyreview.com/s/520426/the-real-privacy-problem/.

Muhlmann, Géraldine. *A Political History of Journalism*. Cambridge and Malden, MA: Polity, 2008.

Muller, Hermann J. *Out of the Night: A Biologist's View of the Future*. London: Victor Golancz, 1936.

——— . 'Genetic Progress by Voluntarily Conducted Germinal Choice'. In *Man and His Future*, edited by Gordon Wolstenholme, 247–62. London: J. & A. Churchill, 1963.

Mutz, Diana C. *Hearing the Other Side: Deliberative versus Participatory Democracy*. New York: Cambridge University Press, 2006.

Nelkin, Dorothy, and M. Susan Lindee. *The DNA Mystique: The Gene as a Cultural Icon*. New York: W. H. Freeman, 1999.

Neues Deutsches Wörterbuch. Köln: Lingen, 2003.

Nord, David Paul. *Communities of Journalism: A History of American Newspapers and Their Readers*. Urbana and Chicago: University of Illinois Press, 2001.

Nowotny, Helga, Peter Scott and Michael Gibbons. *Re-Thinking Science: Knowledge and the Public in an Age of Uncertainty*. Cambridge: Polity, 2001.

Nussbaum, Martha C. *The Fragility of Goodness: Luck and Ethics in Greek Tragedy and Philosophy*. Cambridge: Cambridge University Press, 1986.

O'Banion, John D. *Reorienting Rhetoric: The Dialectic of List and Story*. University Park: Pennsylvania State University Press, 1992.

O'Donnell, Gus, Angus Deaton, Martine Durand, David Halpern and Richard Layard. *Wellbeing and Policy*. Report. London: Legatum Institute, 2014.

Ordbog over det danske sprog. [Dictionary of the Danish language]. København: Gyldendal, 1966 [1919–56].

Ortega y Gasset, José. *The Revolt of the Masses*. New York: Norton, 1993 [1930].

Orwell, George. *Politics and the English Language*. London: Horizon, 1946. Accessed 5 April 2017. http://www.orwell.ru/library/essays/politics/english/e_polit/.

Osborne, Peter. 'Modernity Is a Qualitative, Not a Chronological, Category'. *New Left Review* no. 192 (1992): 65–84. Accessed 5 April 2017. https://quote.ucsd.edu/time/files/2014/03/osborne-qualnotchron.pdf.

Paine, Thomas. *Common Sense*. London: Penguin Books, 2004 [1776].

Philpott, Tom. 'Whatever Happened to Golden Rice?' Genetic literacy project. 4 February 2016. Accessed 2 May 2017. https://www.geneticliteracyproject.org/2016/02/04/whatever-happened-to-golden-rice/.

Pilger, John. *Hidden Agendas*. London, Sydney, Auckland and Parktown: Vintage, 1999.

Popkin, Richard H. *The History of Scepticism from Erasmus to Spinoza*. Berkeley, Los Angeles and London: University of California Press, 1979.

Port Huron Statement. 1962. Accessed 5 April 2017. http://www2.iath.virginia.edu/sixties/HTML_docs/Resources/Primary/Manifestos/SDS_Port_Huron.html.

Porter, Roy. *Enlightenment: Britain and the Creation of the Modern World*. London: Penguin Books, 2001.

Porter, Theodore M. *Trust in Numbers: The Pursuit of Objectivity in Science and Public Life*. Princeton, NJ: Princeton University Press, 1995.

Prabhu, Robinda. 'How Should We Govern the Algorithms That Shape Our Lives?' In *The Next Horizon of Technology Assessment: Proceedings from the PACITA 2015 Conference in Berlin*, edited by Constanze Scherz, Tomáš Michalek, Leonhard Hennen, Lenka Hebáková, Julia Hahn and Stefanie B. Seitz, 237–42. Prague: Technology Centre ASCR, 2015.

Primas, Hans. 'Fascination and Inflation in Science'. In *The Role of Philosophy of Science and Ethics in University Science Education*, edited by Tom Börsen Hansen, 72–90. Helsinki: NSU Press, 2002.

Rathgeb, Eberhard. *Die engagierte Nation: Deutsche Debatten 1945–2005*. München and Wien: Carl Hanser Verlag, 2005.

Redwood, John. *Reason, Ridicule and Religion: The Age of Enlightenment in England 1660–1750*. London: Thames & Hudson, 1976.

Rittel, Horst, and Melvin Webber. 'Dilemmas in a General Theory of Planning'. *Policy Sciences* 4 (1973): 155–69. http://doi.org/10.1007/BF01405730.

Rose, Hillary, and Steven Rose. *Science and Society*. Harmondsworth: Penguin Books, 1971.

Royal Society. *The Public Understanding of Science*. Report of a Royal Society ad hoc group endorsed by the Council of the Royal Society. London: The Royal Society, 1985. Accessed 5 April 2017. https://royalsociety.org/topics-policy/publications/1985/public-understanding-science/.

Sasaki, Masamichi, and Robert M. Marsh, eds. *Trust: Comparative Perspectives*. Leiden and Boston: Brill, 2012.

Schnädelbach, Herbert. *Vernunft*. Stuttgart: Reclam, 2007.

Schneider, Stephen. 'Don't Bet All Environmental Changes Will Be Beneficial'. *APS News Online* 5 no. 8 (1996). Accessed 5 April 2017. http://www.aps.org/publications/apsnews/199608/environmental.cfm.

Schorn-Schütte, Luise. *Konfessionskriege und europäische Expansion: Europa 1500–1648*. München: Verlag C. H. Beck, 2010.

Schudson, Michael. *The Power of News*. Cambridge, MA, and London: Harvard University Press, 2000.

Schwartz, Jason L. 'New Media, Old Messages: Themes in the History of Vaccine Hesitancy and Refusal. *AMA Journal of Ethics* 14, no. 1 (2012): 50–55. Accessed 2 May 2017. http://journalofethics.ama-assn.org/2012/01/mhst1-1201.html.

Schwonke, Martin. *Vom Staatsroman zur Science Fiction: Eine Untersuchung über Geschichte und Funktion der naturwissenschaftlich-technischen Utopie.* Stuttgart: Ferdinand Enke Verlag, 1957.

Sennett, Richard. *The Fall of Public Man.* London: Faber and Faber, 1986.

Siebert, Fred S., Theodore Peterson and Wilbur Schramm. *Four Theories of the Press.* Urbana: University of Illinois Press, 1956.

Sighele, Scipio. *Psychologie des Auflaufs und der Massenverbrechen.* Leipzig and Dresden: Verlag von Carl Reissner, 1897.

Simitis, Spiros. 'Privacy: An Endless Debate'. *California Law Review* 98, no. 6 (2010): 7 [1989]. https://doi.org/10.15779/Z38WM5W.

Simpson, James. *Burning to Read: English Fundamentalism and Its Reformation Opponents.* Cambridge, MA, and London: Belknap Press of Harvard University Press, 2007.

Skloot, Rebecca. 'Under the Skin: A History of the Vaccine Debate Goes Deep but Misses the Drama'. *Columbia Journalism Review* (January/February 2007). Available at https:// archives.cjr.org/review/under_the_skin_a_history_of_th.php

Sloterdijk, Peter. *Die Verachtung der Massen.* Frankfurt am Main: Suhrkamp, 2000.

Smyth, William Henry. *Technocracy: First, Second and Third Series* and *Social Universals.* Reprinted from the Gazette, Berkeley, CA, 1921. Los Angeles: University of California. Accessed 5 April 2017. https://archive.org/details/technocracyfirst00smyt.

Snow, C. P. *The Two Cultures and the Scientific Revolution: The Rede Lecture 1959.* Cambridge: Cambridge University Press, 1959.

Splichal, Slavko. 'Från opinionsstyrd demokrati til globala styrelseformer utan opinion' [From public opinion-based democracy to global forms of governance without public opinion-formation]. In *Demokratirådets rapport 2008: Medierna: folkets röst?,* edited by Olof Petersson, 29–50. Stockholm: SNS Förlag, 2008. Accessed 5 April 2017. http://www. olofpetersson.se/_arkiv/dr/dr_2008.pdf.

Sprat, Thomas. *History of the Royal Society,* edited with critical apparatus by Jackson I. Cope and Harold Whitmore Jones. London: Routledge & Kegan Paul, 1966 [1667].

Stearns, Harold. *America and the Young Intellectual.* London: Forgotten Books, 2012 [1921].

Stehr, Nico. *Knowledge Societies.* London, Thousand Oaks, CA, and New Delhi: Sage Publications, 1994.

Stocking, S. Holly. 'How Journalists Deal with Scientific Uncertainty'. In *Communicating Uncertainty: Media Coverage of New and Controversial Science,* edited by Sharon M. Friedman, Sharon Dunwoody and Carol L. Rogers, 23–41. Mahwah, NJ, and London: Lawrence Erlbaum, 1999.

Strauss, Stefan. '"If I Only Knew Now What I Know Then…": Big Data or Towards Automated Uncertainty?' In *The Next Horizon of Technology Assessment: Proceedings from the PACITA 2015 Conference in Berlin,* edited by Constanze Scherz, Tomáš Michalek, Leonhard Hennen, Lenka Hebáková, Julia Hahn and Stefanie B. Seitz, 233–36. Prague: Technology Centre ASCR, 2015.

Sturgis, Patrick, and Nick Allum. 'Science in Society: Re-evaluating the Deficit Model of Public Attitudes'. *Public Understanding of Science* 13 (2004): 55–74. http://doi.org/ 10.1177/0963662504042690.

Swift, Jonathan. 'The Text of Gulliver's Travels'. In *Gulliver's Travels: An Authoritative Text; The Correspondence of Swift: Pope's Verses on Gulliver's Travels: Critical Essays,* edited by Robert A. Greenberg. New York: W. W. Norton, 1970 [1735].

Szent-Györgyi, Albert. 'The Promise of Medical Science'. In *Man and His Future*, edited by Gordon Wolstenholme, 188–95. London: J. & A. Churchill, 1963.

Tarde, Gabriel. *The Laws of Imitation*. New York: Henry Holt, 1903.

Tavernor, Robert. *Smoot's Ear: The Measure of Humanity*. New Haven, CT, and London: Yale University Press, 2007.

Tomalin, Claire. *Samuel Pepys: The Unequalled Self*. London: Penguin Books, 2003.

Toulmin, Stephen. *Cosmopolis: The Hidden Agenda of Modernity*. New York: Free Press, 1990.

Underwood, Doug. *From Yahweh to Yahoo! The Religious Roots of the Secular Press*. Urbana and Chicago: University of Illinois Press, 2008.

Veblen, Thorstein. *The Theory of the Leisure Class*. 1899. Accessed 2 May 2017. http://xroads.virginia.edu/~HYPER/VEBLEN/veb_toc.html.

Verne, Jules. *The Lost Novel: Paris in the Twentieth Century*. New York and Toronto: Ballantine Books, 1996.

Ward, Bob. 'Reporters Feel the Heat over Climate Change'. *The Independent*, 10 March 2008. http://www.independent.co.uk/news/media/reporters-feel-the-heat-over-climate-change-793586.html.

Ward, Lester Frank. 'Sociocracy' (slightly abbreviated text from: Lester Frank Ward, *The Psychic Factors of Civilization*, chap. 38, 313–30. Boston, MA: Ginn, 1893). Accessed 21 April 2017. http://www.slideshare.net/dinugherman/sociocracy.

Weber, Max. *Politik als Beruf*. Stuttgart: Reclam, 1992 [1919].

Wells, H. G., Julian Huxley and G. P. Wells. *The Science of Life: A Summary of Contemporary Knowledge about Life and Its Possibilities*. Part 1–31. London: Amalgamated Press, 1929–30.

Welsh, Ian, and Brian Wynne. 'Science, Scientism and Imaginaries of Publics in the UK: Passive Objects, Incipient Threats'. *Science as Culture* 22, no. 4 (2013): 540–66. http://doi.org/10.1080/14636778.2013.764072.

Williams, Bernard. *Ethics and the Limits of Philosophy*. London: Fontana Press, 1993 [1985].

Wolstenholme, Gordon, ed. *Man and His Future*. A Ciba Foundation Volume. London: J. & A. Churchill, 1963.

Wood, Gordon S. *The Radicalism of the American Revolution*. New York: Vintage Books, 1993.

Worden, Blair. *The English Civil Wars, 1640–1660*. London: Weidenfeld & Nicolson, 2009.

Wright, C. Mills. *The Power Elite*. New York: Oxford University Press, 2000 [1956].

Zenker, Frank. 'The Explanatory Value of Cognitive Asymmetries in Policy Controversies'. In *Between Scientists and Citizens: Proceedings of a Conference at Iowa State University June 1–2, 2012*, edited by Jean Goodwin, 441–51. Ames, IA: Great Plains Society for the Study of Argumentation, 2012.

Ziman, John. *Real Science: What It Is, and What It Means*. Cambridge: Cambridge University Press, 2000.

———. *Reliable Knowledge. An Exploration of the Grounds for Belief in Science*. Cambridge, New York and Melbourne: Cambridge University Press, 1978.

INDEX

Lightning Source UK Ltd.
Milton Keynes UK
UKHW03n0137020418
320358UK00001BA/9/P